American Samoa and the Swains Island Expeditions

A Tale of Ancient Culture, Modern Science, and a Marine Sanctuary in the Pacific

Daniel J. Basta

DORRANCE PUBLISHING CO
EST. 1920
PITTSBURGH, PENNSYLVANIA 15238

Dorrance Publishing Co
585 Alpha Drive
Pittsburgh, PA 15238
Visit our website at www.dorrancebookstore.com

ISBN: 979-8-89127-934-6
eISBN: 979-8-89127-432-7

American Samoa and the Swains Island Expeditions

*A Tale of Ancient Culture, Modern Science, and
a Marine Sanctuary in the Pacific*

Daniel J. Basta

Dedication

To Fa'a Samoa and the Samoan People

Preface

When I began to write this book, the story I had in mind was to tell the tale of two expeditions to Swains Island in American Samoa in which I had participated. However, I soon found myself writing something a little different: an important story to illuminate was how circumstances projected American Samoa onto the international stage, in terms of marine protected areas in the Pacific.

Therefore, the book evolved along three lines or threads of storytelling, all within the context of telling the stories of the expeditions. The first was to write about American Samoa, some of the people I have known, and how I have come to see this place. The second was to describe how the National Marine Sanctuary of American Samoa came about between the First and Second expeditions; the creation of which I, along with many others, had something to do with. I describe this thread in italics to distinguish it from the primary story. Third was to describe the tale of the two expeditions within the context of the first two threads.

Told together in this way, the threads may be more informative than each told separately. This may have led to a little choppiness in the telling. The reader will have to be the judge of that. The story unfolds when an expedition to Swains Island is a figment of imagination and concludes with a second expedition and its aftermath. All is essentially told through the lens of the expeditions, revealing that everything in this small island archipelago is always connected.

Most importantly, the overall story reveals how one decision leads to another, ultimately creating a path resulting in the largest expansion of a marine sanctuary or protected area in the history of the United States. If there is a lesson in this story for others struggling in the world of marine conservation, it may be to broaden their perspective on how to consider the little things that they do that, when taken together, can make a difference – a big difference.

Acknowledgments

As I wrote this book, I came to the realization that I owed much to many more individuals than just those who had reviewed the manuscript. This may often be the case when writing about firsthand experiences that take place over time and in which so many individuals were involved in subtle ways. Many of these individuals may not even realize they played a role. Their influences often took place through random conversations in the corridor, over drinks after work, by sharing adjoining seats on an airplane, and even at meetings that had nothing to do with American Samoa and Swains Island. These unsung colleagues and participants were the sounding boards that helped form my approach in American Samoa as well as the creation of the expeditions to Swains Island. They are far too numerous to name, and naming anyone would undoubtedly leave someone out. I want to thank all of them, however, for letting me intrude into their "space" when it moved me. They were the best part of the open and collaborative approach we used in those days.

Returning to the story itself, I must acknowledge specific individuals. I especially want to thank my good friend Togiola Tulafono, the former governor of American Samoa, who not only read the manuscript and provided valuable commentary, but who also was a guiding light throughout all my endeavors in American Samoa.

Thank you to Gene Brighouse, who instantly responded to my many and varied questions about events, timing, and individuals, as well as providing me with valuable background materials. To Bill Kiene, who early on provided background information and also carefully reviewed portions of the manuscript. To Hans Van Tilburg, who provided background materials and pictures from both his original report and the science report on the Second Expedition; he also gave a careful review of the entire manuscript, which helped me get it right. To Jim Knowlton, who reprised his role for me on the Second Expedition and got it right in working with Jean Michel Cousteau on the documentary film. To Paul Chetirkin and Dana Wilkes, who clarified some of the events in the story and encouraged me to write this book. To David Jennings and Alex Jennings, who read the manuscript and provided encouragement to add to the legacy of their beloved Swains Island. To Brad Barr, who read some of the early chapters and raised my awareness of cultural sensitivities. To my friends, Dave Hanley and Larry Northup, to whom I told stories many times. To my editor, Ann Boese, who edited the manuscript and helped me find my voice in all things I write. To Pam Rubin, a great editor herself, who proofread the final manuscript.

Lastly, to my wife, Arlene, who accompanied me to American Samoa in 2009 and got a feel for the place, its people, and my special relationship with them. She encouraged me in all of my endeavors and patiently indulged my many long absences while I was in American Samoa and the South Pacific. I think I missed five birthdays in a row while working in American Samoa and journeying to Swains Island. I could not have done any of this without her support.

Table of Contents

List of Maps and Photographs
Photographs are found at the end of each chapter

Reference Maps

Map 1. Pacific Ocean Location of Samoan Archipelago.

Map 2. The Samoan Islands archipelago.

Map 3. The Samoan Islands and the U.S. Territory of American Samoa.

Map 4. "Tiny" Fagatele Bay off Tutuila.

Maps 5 & 6. Satellite images of remote Swains Island.

Maps 7 & 8. Swains Island maps.

Map 9. The National Marine Sanctuary of American Samoa.

Map 10. The special sanctuary protected areas at Swains Island.

Map 11. The island of Aunu'u.

Photographs

Chapter 3

Picture 3.1. Looking out onto "tiny" Fagatele Bay.

Picture 3.2. [**Waiting for Picture**]

Picture 3.3. The persistent Alex Jennings.

Reference Maps

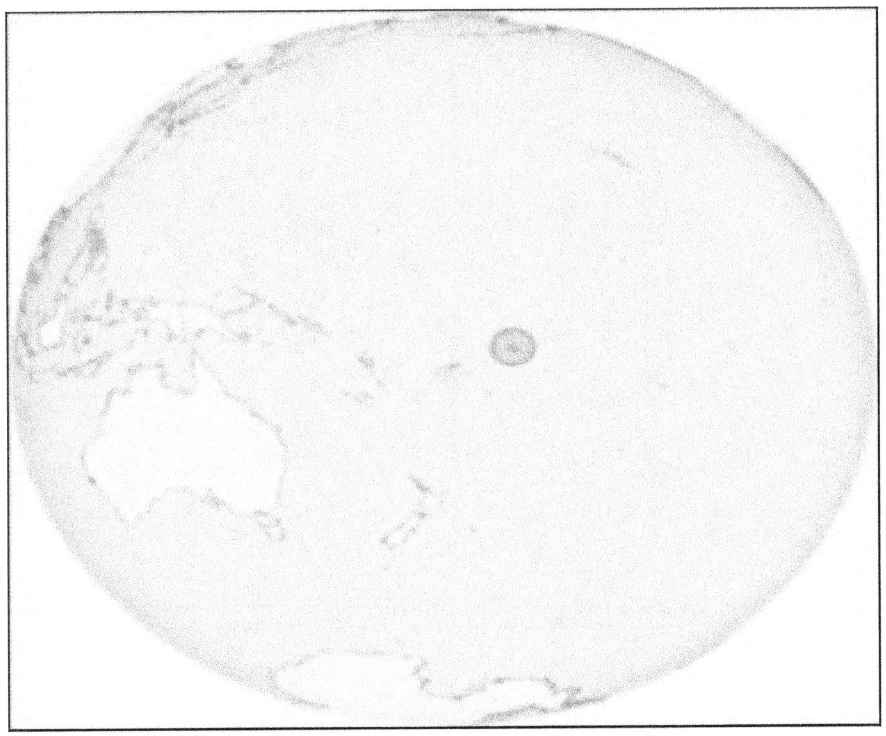

Map 1. The Pacific Ocean and location of the Samoan Islands archipelago.

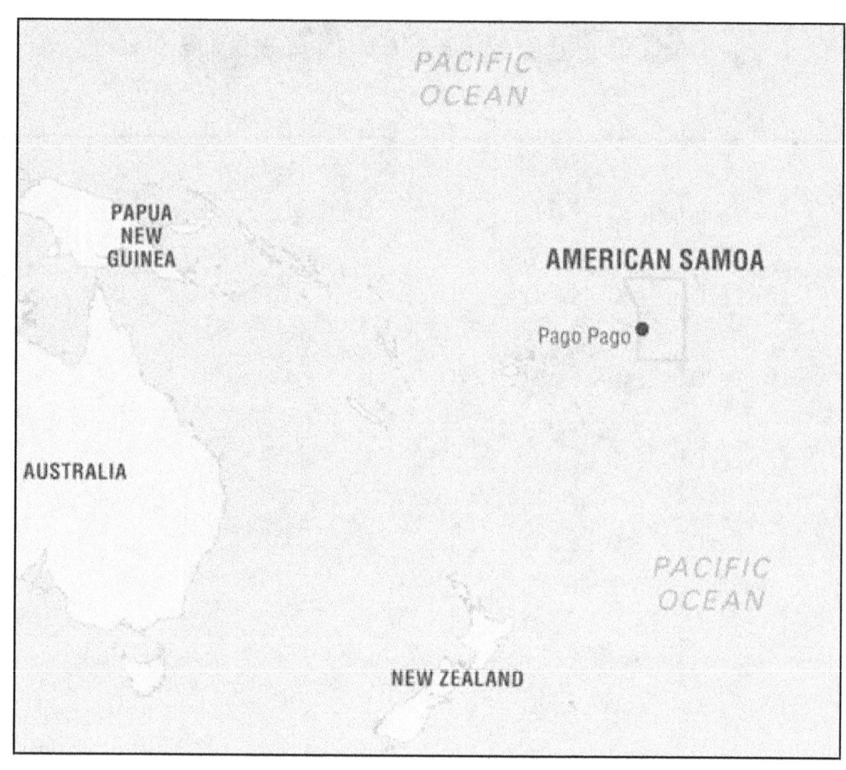

Map 2. Location of American Samoa in the South Pacific.

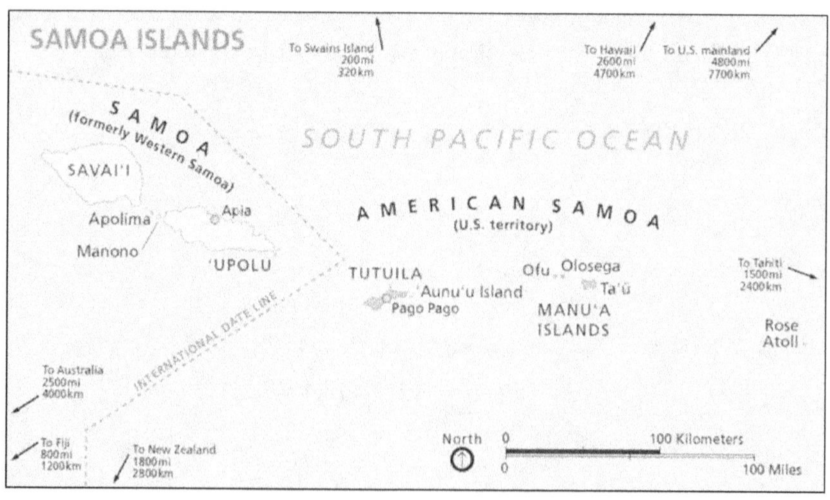

Map 3. The Samoan Islands and the U.S. Territory of American Samoa. Note the reference to Swains Island, *at top*.

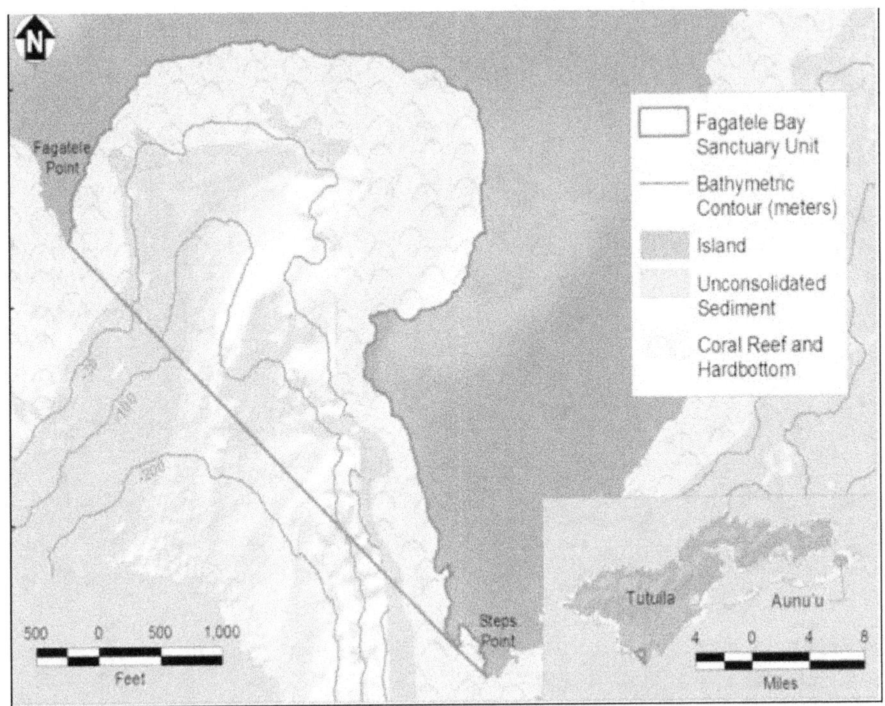

Map 4. Tiny Fagatele Bay. The Fagatele Bay Unit, *above the diagonal line,* is within the National Marine Sanctuary of American Samoa.

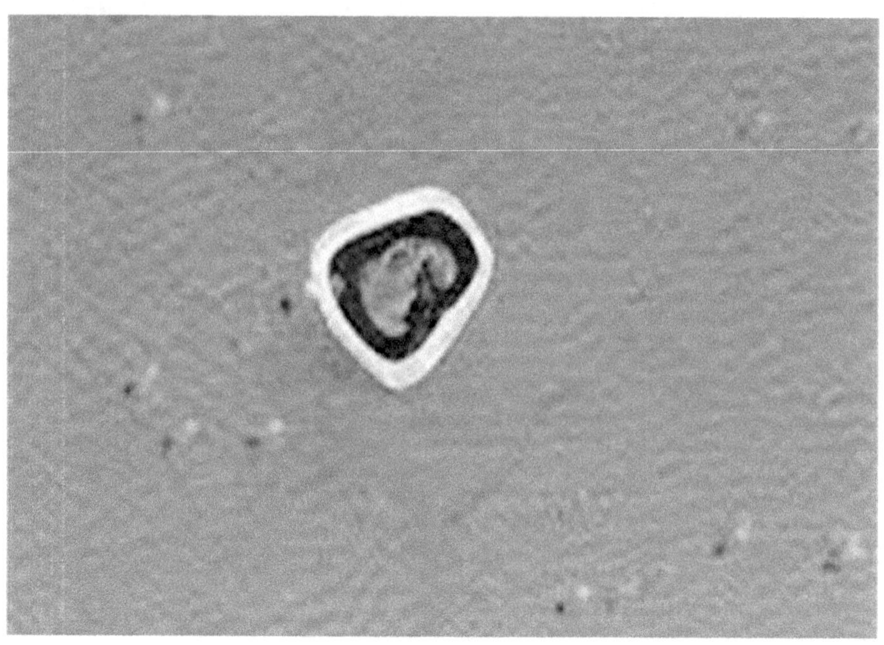

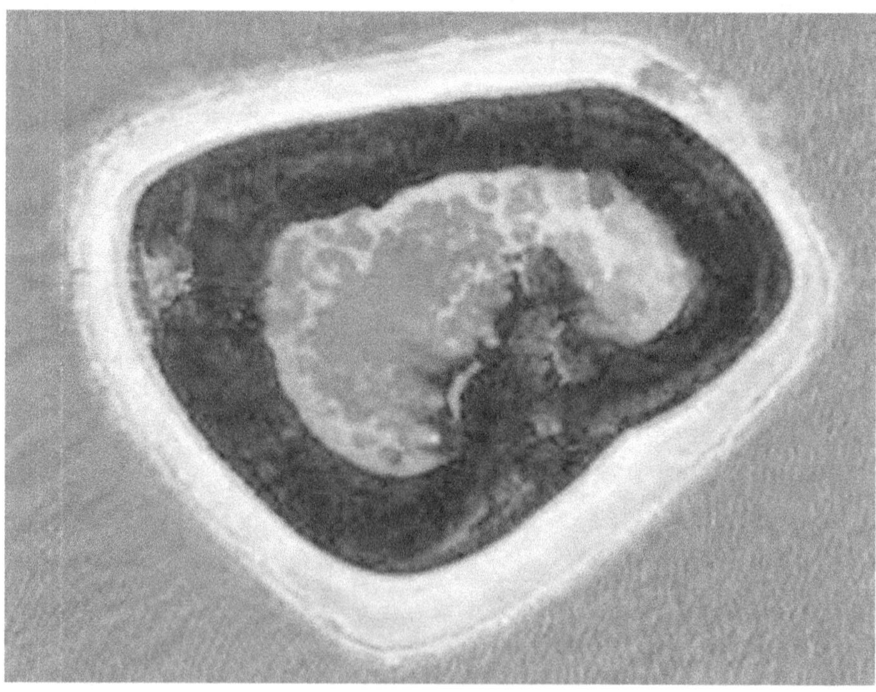

Maps 5 & 6. Satellite images of remote Swains Island.

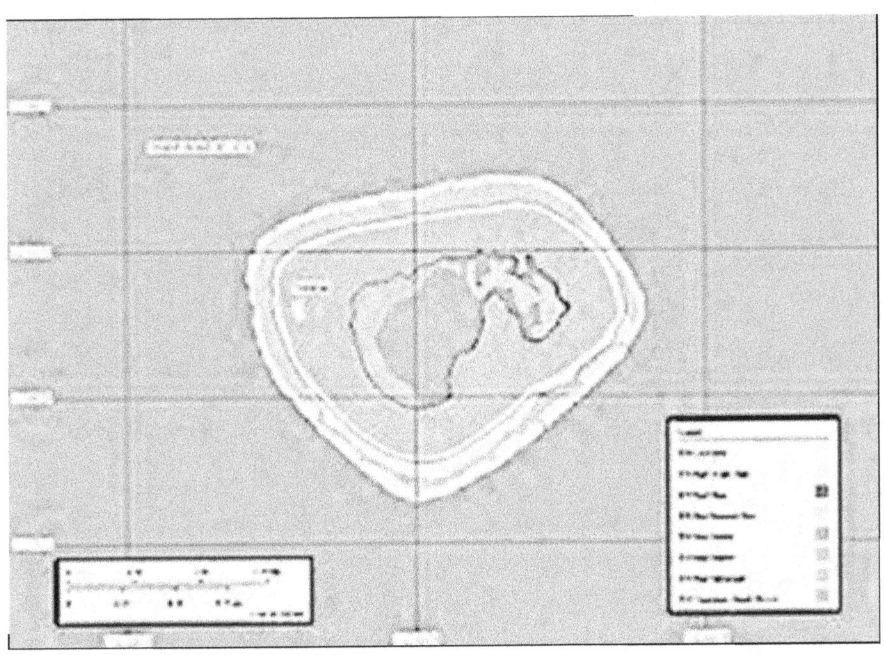

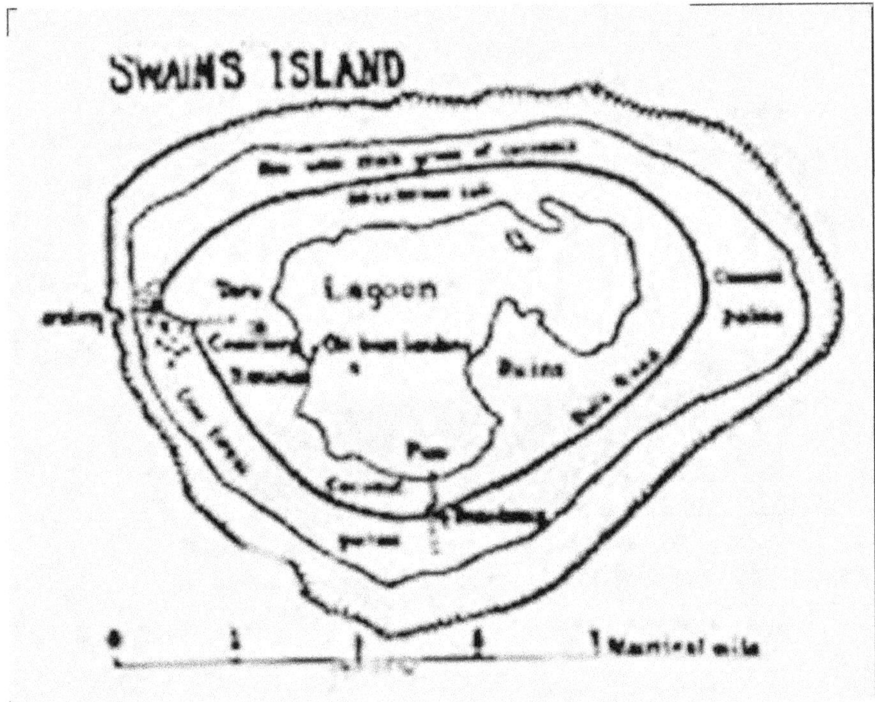

Maps 7 & 8. Swains Island maps.

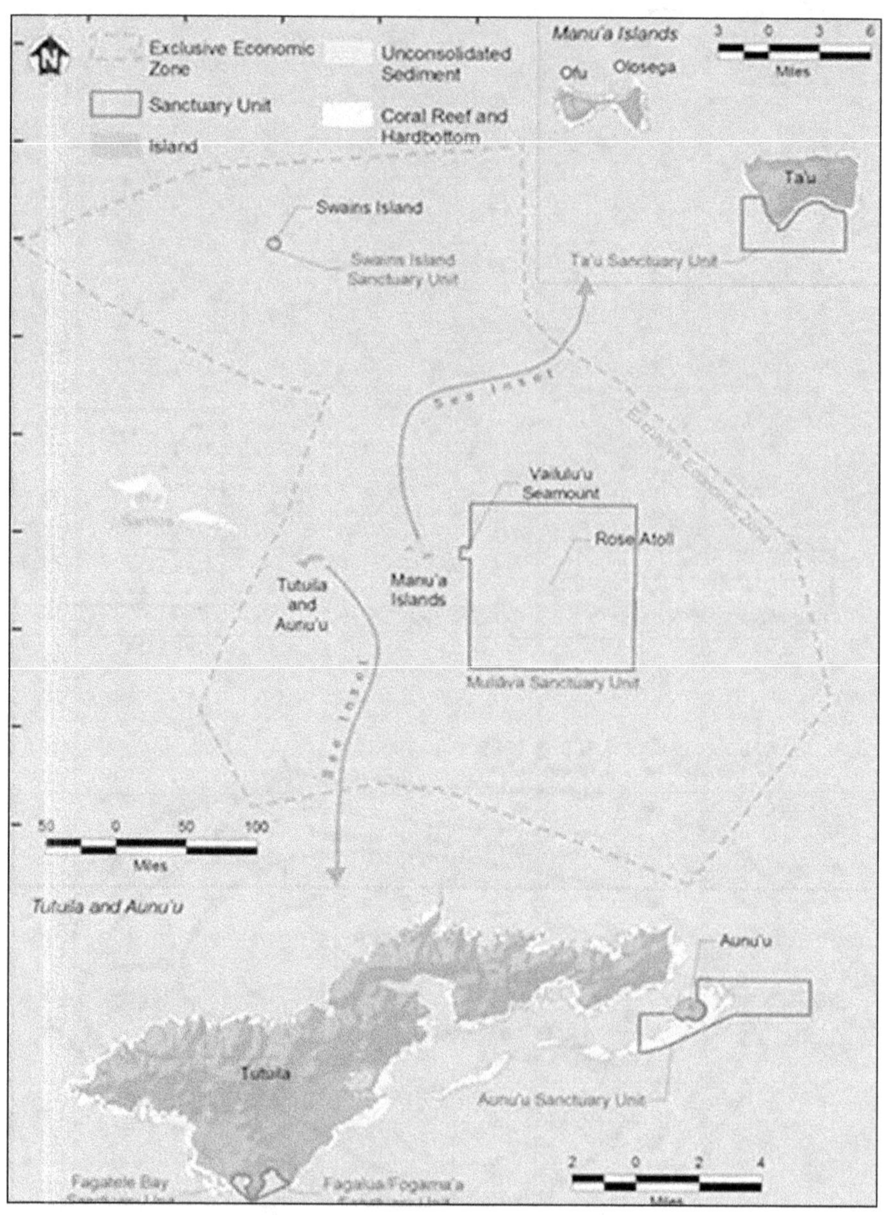

Map 9. The National Marine Sanctuary of American Samoa.

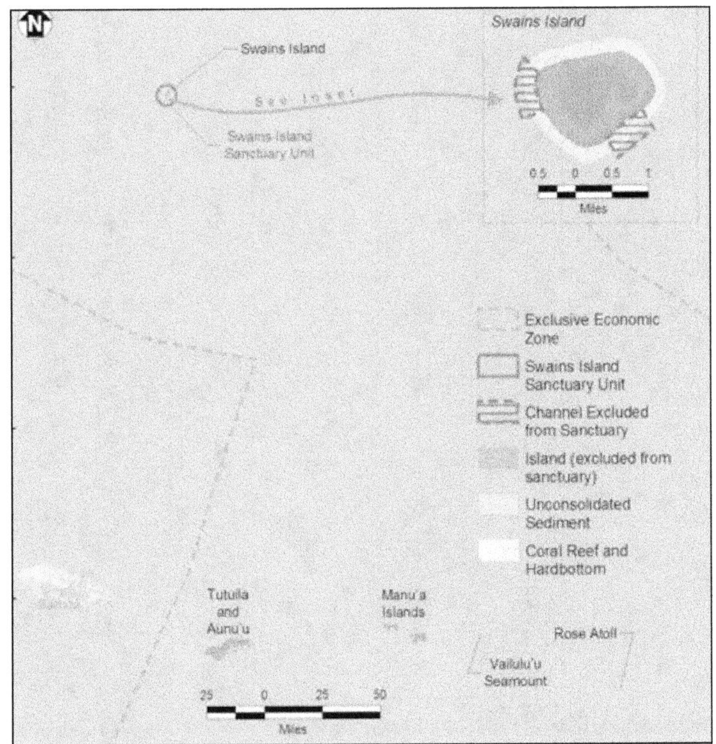

Map 10. The special sanctuary protected areas at Swains Island.

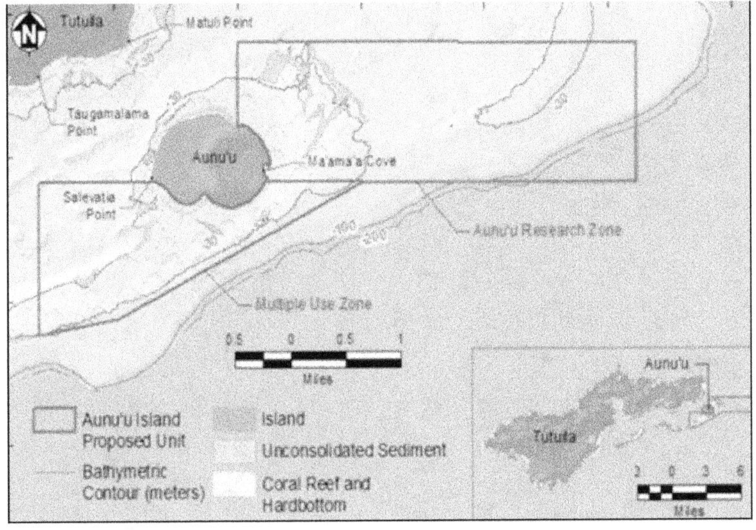

Map 11. The island of Aunu'u.

Chapter 1

Adventures in a Faraway Place

This is a book about the adventure and story behind two expeditions to a tiny, remote, unique, and currently uninhabited island in the South Pacific. Swains Island (also known as Olosega), lies over 200 miles due north from Pago Pago in the U.S. Territory of American Samoa (*Maps 5, 6*). There is no airport or harbor at Swains Island. It is hard to get to. It is even harder to get through the surrounding narrow reef and surf to get ashore. The island covers an area of only about 1.3 square miles and was last inhabited in 2008. The story told here is as much about American Samoa and how two expeditions came to be as it is about the expeditions themselves. For each expedition, I have tried to describe both the circumstances and events leading to them, the interesting people involved, and their roles as events came together in unexpected ways — at least from my perspective.

The First Expedition was undertaken in 2011 (*Chapters 5-7*) and ultimately drove my curiosity, which led to a Second Expedition in 2013 (*Chapters 8-12*). The primary reason for the Second Expedition was to search for artifacts — potentially left behind by ancient Polynesian voyagers — that now might be entombed in the brackish lagoon in the center of Swains Island. In the telling of these expeditions, I reveal the history of Swains Island, its ecology, and its secrets; describe how the expeditions supported the eventual creation of the extraordinary National Marine Sanctuary of American Samoa in 2012; and much about American Samoa and its people.

Early in the first discussions of an expedition to Swains Island, I had been fortunate to convince my good friend, Jean Michel Cousteau, that Swains Island was a unique and remote Pacific atoll that he should know about. Jean Michel is an old hand in the Pacific but had never heard of this place called Swains Island. My pitch to him was that his involvement with Swains Island could provide an opportunity to communicate about ways in which small Pacific island ecosystems could be sustained and how sustainable tourism might be created for such places. The idea of potentially discovering artifacts from the great age of Polynesian voyaging came later, during the First Expedition. More than anything else, it was unraveling the mysteries in the lagoon that sparked my interest in launching the Second Expedition.

Jean Michel's initial involvement helped to spur on the First Expedition. He was all set to accompany me and the others on the expedition, when fate intervened a week before the ship was to leave Pago Pago, and he never made it. Two years later, however, we sailed on a Second Expedition, with a cobbled together science party and others, through a harrowing overnight transit to Swains Island. Accompanying us was Jim Knowlton, a filmmaker with a long history of making documentaries with Jean-Michel and others. They would later produce an award-winning documentary film of the Second Expedition: *Swains Island — One of the Last Jewels of the Planet*. To most who have seen it, this film stands out as the basis for what they know of Swains Island. It is also a record of the participants and the science party, spoken in their own words. Although an excellent scientific report of the Second Expedition was written, *Unlocking the Secrets of Swains Island: A Maritime Heritage Resources Survey*, from which I have drawn much, such reports are not widely distributed and remain primarily of interest to scientists.

Consequently, I have chosen to write this book to cover a broader history of the expeditions and American Samoa than either the film or science report have captured. *In addition, a thread runs through the book about the evolution of the tiny Fagatele Bay National Marine Sanctuary and its unprecedented expansion from .25 square miles to the National Marine Sanctuary of American Samoa, encompassing 13,581 square miles.* The expeditions to Swains Island were part of a larger set of activities going on in American Samoa that drove the creation of the enormously expanded "new" marine sanctuary between the first and second expeditions.

I have attempted to capture the people, their circumstances and relationships, and the serendipity of events, which became the mixing bowl that gave birth to the expeditions and the expanded Sanctuary. I think it is the type of story that lurks behind the scenes of many events and is often never told. In fact, the journey to set foot on Swains Island is as important as the expeditions themselves. Not only is the journey of interest in its own right, but it is also instructive on many levels. It provides insights into how little things converge and illustrates how anyone can put such things in motion. In some sense, this bigger story reveals a "how to make it happen" journey. I would like to think others will find inspiration and see how they, too, can create and navigate such endeavors to add value to the places and causes they care about.

American Samoa is a small place in a very big ocean (*Map 1*). The Polynesian communities and families who live there are as culturally intact as exist anywhere in the world today. Serendipity made it my great fortune to go there and, over time, to be adopted into this family of Polynesians. They taught me many things so the telling of their story has been particularly important for me to do.

The story is, of course, my interpretation based on recollections, personal notes, research materials, and discussions with some who were there. I think this book connects many "dots" about which even the participants still may not be aware. And so, this book is my personal account and a partial memoir of my experiences in American Samoa. I hope that the story is interesting, exciting at times, inspirational, and, most of all, a good read. Any omissions, faulty memories from the passage of time, and errors in the final telling are mine. The narrative is probably best described by the popular phrase, "As I remember it." Even so, I think I got the story mostly right.

Chapter 2

A Short History of American Samoa and Swains Island

American Samoa is part of the Samoan Islands archipelago (*Maps 1 and 2*) and is comprised of only the smaller eastern portion of the archipelago. The Samoan Islands are generally steep and mountainous and were formed along a tectonic plate, much like the Hawaiian Islands. The archipelago (*Map 3*) stretches along the 14th degree longitude line south latitude for almost 300 miles, from Rose Island and the Manu'a group of small islands in the east to the much larger islands of Upolu and Savai'i in the west. Together with the "nearby" islands of Tonga — and, to some extent, Fiji — Samoa is considered one of the cradles of the great Polynesian voyaging culture that crossed and inhabited the seemingly limitless Pacific.

Polynesian Voyagers. The ancestors of the Polynesian voyaging culture reached Samoa around 3,000 years ago and continued their voyages east into the sun, reaching Hawaii perhaps 1,200 to 700 years ago. It is known that these great navigators made voyages not only to the east but also back again to the west. Periodic voyaging among the islands of the Pacific is also known to have taken place. The full extent of Polynesian voyaging, however, remains lost in the veils of time. Many references are now available describing this — the greatest of all human migrations — and further research is ongoing. There are two books about Samoa that I especially value, which tell the history and story of the Samoan Islands and American Samoa. They are both by J. Robert Shaffer: *American Samoa: 100 Years Under the United States Flag (Centennial Edition)* (Honolulu: Island Heritage Publishing, 2000) and *Samoa: A Historical Novel* (New York: Ithaca Press, 2011). They are so well done, and, together, they tell a complete and fascinating story. Read them and you shall know more than you may ever want to know about Samoa and Samoans.

Coming of the Europeans. It wasn't until the late 1700s that Europeans really became interested in the Samoan archipelago and that region of the South Pacific. There is evidence, however, that European explorers had reached the islands in the South Pacific much earlier. In 1609, for example, the explorer Pedro Fernandez de Quiros described landing on a small island inhabited by a light-skinned and peaceful people in the South Pacific. In the early 1800s, whaling ships from the west began to journey into the South Pacific in search of their prey. Following in their wake were the missionaries and, eventually, others. By the middle of the nineteenth century, the European powers had begun to take a serious economic interest across the entire Pacific. This "interest" led to an age of dominating colonization, imperialism, and, eventually, the "annexations" that would engulf the entire Pacific.

American colonial interest in the Pacific lagged behind that of the other powers and hadn't really taken hold until the end of the nineteenth century, following the Spanish-American War in 1898. Nonetheless, American whaling and merchant ships were plying the Pacific in increasing numbers throughout the era. As mid-century approached, U.S. maritime and commercial interests prompted the United States Exploring Expedition, from 1838 to 1842, to explore and survey the Pacific Ocean and its surrounding lands. Often referred to as the "Wilkes Expedition," for the naval officer who commanded it, this expedition led to the first contact between the American government and the islands of Samoa. Wilkes and his party visited the Samoan Islands of Upolu in the west and Tutuila in the east. The expedition was the first to survey and chart the harbor at Pago Pago on Tutuila, which ultimately played a role in the annexation of what became known as "American Samoa."

European Domination. Like all indigenous islanders in the Pacific, the Samoans fought vigorously against domination by the competing European powers, primarily England and Germany, in the South Pacific. Nevertheless, domination progressed inevitably through the arrival of missionaries and the economic transformation of the islands in the Pacific to serve European and American markets. The creation of *copra* (dried coconut), cotton, and sugar cane economies — and, ultimately, military force — had overwhelmed Samoa by the end of the nineteenth century. The Samoans, however, maintained their traditions and their culture through it all. Although they were dominated in many ways, the Samoans had never given in.

I think it is fair to describe this period in the Pacific as chaotic, at best. In a real sense it was like the proverbial Wild West. Nations, corporations, and individuals manipulated inter-island and tribal rivalries and took whatever they wanted. The only common denominator was that the Polynesians, Melanesians, and Micronesians always lost out. It was a dark period for the peoples of the Pacific.

By the late 1880s, Germany had become the dominant European power in control of Samoa, much as they were becoming elsewhere in the south and central Pacific. American and German interests in Samoa, however, were coming into conflict, particularly the American interest in the excellent harbor at Pago Pago on Tutuila. In 1889, the conflict came perilously close to igniting a war: German and American warships faced each other eyeball-to-eyeball in a tense situation in Apia harbor on the island of Upolu, just to the west of Tutuila and the Pago Pago anchorage. Fingers were on triggers when Mother Nature intervened, and a sudden horrific typhoon struck Apia. The impartial typhoon sank three American and three German warships, with a great loss of life. The typhoon "dampened" the moment of crisis, like throwing water on two alley cats about to grapple.

American Samoa Is Created. Over the next decade, sane minds prevailed. After much maneuvering on all sides, the islands of Samoa, which had been home to its people for almost 3,000 years, were divided east from west, as were the Samoan people who inhabited them. Germany maintained "ownership" of the larger western islands, Upolu and Savai'i, and most of the population, while ceding authority over Tutuila and the smaller islands to the east to the United States. In 1900, the United States formally annexed American Samoa as a U.S. territory, as it remains to this day. With the defeat of Germany at the end of World War I, German or Western Samoa became a British protectorate governed by New Zealand. In 1962, Western Samoa became the first independent "microstate" in the Pacific, and, in 1970, joined the British Commonwealth. In 1997, Western Samoa officially altered its name simply to "Samoa," as it is known today.

As for American Samoa, the territory was governed by a long list of U.S. naval officers appointed as administrators from its annexation in 1900 until 1951. After 1951, the administration of American Samoa was finally passed to U.S. appointed civilian governors. Then, in 1978, administration was passed to elected Samoan governors. Samoan governors are now elected every four years and can be elected to a second term. An elected *ex officio* representative of American Samoa also sits in the U.S. Congress, representing the people of American Samoa. This is how it was when I first came to American Samoa and began a journey that led to the Swains Island expeditions.

Swains Island Saga. Polynesian voyagers are believed to have visited tiny Swains Island (*Maps 5-8*), also known as Olosega in the local dialect, and even to have settled there in the distant past. Although small, the island had sources of fresh water, abundant vegetation, hardwoods, and seabirds, and was surrounded by waters teeming with life. It still has many of these qualities today. Its small size and remote location, however, always limited the number of inhabitants it could support. Oral histories indicate that, at various times, the island was dominated by inhabitants from larger neighboring atolls, such as those from the Tokelauan Islands and possibly even from more distant Samoa itself. There is also a belief that Olosega may have been used as a waypoint by Polynesian voyagers traveling from Samoa to other distant islands. One thing is clear: the distant past of Swains Island will remain forever murky. The Polynesians left no written records, and much of their oral history has been lost over time. This sometimes difficult process of archeology in the Pacific continues to this day. Who knows what may be rediscovered about the Polynesians — and even the distant past of Swains Island?

When western whalers began to enter the South Pacific in the early 1800s, they also began to make calls at Swains Island. It was an ideal place to reprovision. It was reported that, in 1841, a whaler from New England passed on the position of the island to the Wilkes Expedition so it could be placed on American charts. How many whalers had stopped at Swains Island is unknown. It is known that, in the 1840s, a copra plantation was documented to have been established on the island, presumably by several Frenchmen. Swains Island was not immune to the naked colonialism and unbridled capitalism that dominated the Pacific then. In fact, its very small size and remoteness made it extremely vulnerable. It wasn't long after contact that many of the indigenous inhabitants of the island, a Tokelau people, had fled. It appears that no one spoke for tiny Swains Island at that time. It was just another faraway place engulfed in the Wild West ethos sweeping the Pacific.

In 1856, Eli Jennings, Sr., a whaler and shipwright from the town of South Hampton on eastern Long Island, New York, arrived at Swains Island with his Samoan wife, Malia, from Upolu. They purchased the title to the island from an Englishman named Captain Turnbull, who claimed to have discovered the island. It was the same old story told since the first step taken ashore by Christopher Columbus in the New World. In the 1800s, Europeans still felt entitled to claim ownership of any place where they could get away with it. The Pacific became the last frontier for them to lay claims.

At the time Eli Jennings purchased Swains Island, a few Tokelau people may still have been living there. Others also claimed ownership. By the 1880s, however, Jennings had established a successful copra plantation and registered tiny Swains Island as a private property with the American Consulate on

Upolu. Eli Jennings, having been a whaler, also tried to establish a whaling station on the island. Although whaling around the island quickly waned, copra continued to be exported from the island — even into the 1960s — and supported a small, dwindling population. In 2008, the last inhabitants left Swains Island.

There had been a continuing and long debate about who had sovereignty over Swains Island and how it was to be determined. Was it Western Samoa (first German, then British), American Samoa, or Tokelau (German, then British)? At the end of WWI, sovereignty across the Pacific began to be clarified, and it was more complicated than would meet the eye. The American annexation of Swains Island in 1925, under the authority of the U.S. Naval Administration in American Samoa, settled the issue of sovereignty for all intents and purposes. It wasn't until 1981, however, that New Zealand, of which Tokelau is still a dependent territory, confirmed U.S. sovereignty over Swains Island. One still hears murmurings about that decision, and I have heard them more than once. Nevertheless, Swains Island is part of American Samoa, and is, therefore, a U.S. territory in the South Pacific.

Alex Jennings, who had grabbed my ear in 2005, is the great-grandson of Eli Jennings. On behalf of Swains Island, the Jennings family descendants receive an annual sum from the Territory of American Samoa to allow them periodically to visit their island and connect to their heritage. This connection would eventually underlie the expeditions to Swains Island.

Chapter 3

In the Beginning

To be clear: From the earliest glimmer of an idea, an expedition to Swains Island had to have purpose within a larger framework in American Samoa to add value to the goal of sustainable development and stewardship of the islands and the lives of the Samoans who inhabited them. If the tiny Fagatele Bay National Marine Sanctuary was to have some purpose in the scheme of things, it, too, had to project how it supported this goal. It followed that an expedition to Swains Island had to be a means toward this goal, and not an end in and of itself.

Background. By the early 2000s, I had been traveling around the vast Pacific for some time. I had even passed through American Samoa once on a flight to Australia in the 1970s, but that was only for a couple of hours. I remember it was on the afternoon of a bright sunny day, and it had been refreshing to breathe the sea air and stretch my legs. I could smell and hear the sounds of the island of Tutuila but could not see anything outside of the airfield along the sea. In those days, mostly DC-10s flew across the Pacific, and they always had to island-hop to stop for fuel. I had come to know a little about Polynesian culture from visiting Tonga and Fiji, among other places in the South Pacific. I also picked up some knowledge on many visits to Hawaii. I knew very little of Samoa, however, and my knowledge was confined to what the tourist books said.

All of this began to change in 1999, when I was asked to take over the U.S. National Marine Sanctuary System, administered by the National Oceanic and Atmospheric Administration (NOAA) within the U.S. Department of Commerce. I had already been in NOAA for 20 years, but at that time had only a passing relationship with National Marine Sanctuaries. My only experience came from periodically applying the science and management expertise of the division I ran within the National Ocean Service in NOAA to help a few of the fledging National Marine Sanctuaries through their designation process and to help resolve some of their assessment and management problems. National Marine Sanctuaries was a very small program buried deep within NOAA, and it was always short on resources and expertise. They needed all the help they could get. I have come to conclude that this will always be the case for such programs in government.

It came as a complete surprise when Nancy Foster, the Assistant Administrator of NOAA's National Ocean Service, asked me to consider taking over the struggling National Marine Sanctuary System. I had never even given it a thought. I was comfortable in the science, analysis, and assessment world that I had created in my corner of NOAA. Apparently, Nancy knew something about me that I didn't. Little did I know then that my decision to accept her offer would begin the road to American Samoa and the Swains Island expeditions that followed. My personal interests and expeditions into the Pacific were about to align with my new responsibilities in NOAA. Perhaps it was kismet, but it had surely been Nancy Foster's doing.

In 1999, two National Marine Sanctuaries were in the Pacific, not including those on the U.S. West Coast bordering this great ocean. One was the Humpback Whale Sanctuary in waters surrounding the main Hawaiian Islands, and the other was the tiny sanctuary of Fagatele Bay off the island of Tutuila in American Samoa (*Map 4 and Picture 3.1*). At first, I could not grasp the importance of the Fagatele Bay National Marine Sanctuary in the overall scheme of things. It was only 0.25 square nautical miles in size and located in the middle of the Pacific. For everyone in government, Fagatele Bay was out of sight and out of mind. It seemed to me to be a "curio" of sorts, maybe a placeholder for something in the mind of a politician or two. *There was neither discussion nor even a hint of expanding this tiny place.* It could be said that the Sanctuary was barely hanging on.

The Sanctuary in American Samoa. Nancy Daschbach, the Fagatele Bay Sanctuary manager, had been living and working in American Samoa for many years. Nancy had done her best with the very meager resources allocated to the site. She certainly knew a thing or two about American Samoa and was able to build a small yet active education and research program for tiny Fagatele Bay. Nancy had previously worked in the American Samoan territorial system and subsequently made sure Fagatele Bay was always in the "conversation" in American Samoa. Nancy, a biologist from West Virginia, had fallen in love with Samoa and was raising her family on Tutuila. She taught me a lot during my early visits, especially about beginning to understand the "Samoan way." This continued through 2006, when she left Samoa and moved to Hawaii. As of the start of writing this story, she still worked for the Office of National Marine Sanctuaries in Hawaii. This was convenient for me because I could still get her advice when I needed it.

As is the case in so many indigenous places, the "Western face of government" in American Samoa often masks, to most outsiders, the way in which the culture actually works and does business. Early on, I had a lot to learn about American Samoa and Samoans (and probably still do). I didn't hear about Swains Island for some time. It was the confluence of my ideas about how a National Marine Sanctuary must be part of a community, certain personalities in American Samoa, and my own interests, that eventually led to the expeditions to Swains Island.

The Citizens Advisory Council. One of the things that I brought to all marine sanctuaries was the formation of "citizen advisory councils." It had taken a while to get these councils established and then become an integral part of how

U.S. National Marine Sanctuaries are managed. This aspect of government —
involving the public in the management of such places — is now mostly taken
for granted. It wasn't back then, and it was actually a struggle to bring this
concept to fruition.

Citizen advisory councils became a critical partnership on my "watch" in all
Sanctuaries to ensure that our programs were actually supporting communities
and getting done what mattered to local people. One thing they also did was to
connect me to individuals from all walks of life in a location. I made it a point
to be personally involved in these councils. I guess it might have just been my
style to do it this way. The individuals who voluntarily served on the councils
are among the most committed citizens I have ever met. It wasn't until 2005,
however, that an advisory council could be established in American Samoa for
Fagatele Bay. Establishing a citizen advisory council in American Samoa was a
different proposition than it was on the mainland. It had to be created in the
"Samoan way" with Samoans; thus, Nancy Daschbach had her hands full. It
was at an early advisory council meeting in Pago Pago that the real journey
began that led to the expeditions to Swains Island.

First Rumors of Swains Island. It was at one of these early meetings of the
Fagatele Bay advisory council (*Picture 3.2*) that I first heard of Swains Island.
Alex Jennings was a member of the sanctuary advisory council and also the
representative of Swains Island to the *Fono*, or Territorial Assembly of American
Samoa (*Picture 3.3*). He is a passionate spokesman for Swains Island, which
came under the ownership of his great-grandparents, the Jennings family,
during the nineteenth century. Although now uninhabited, the island was still
owned by the family. Listening to Alex, I discovered that Swains Island seemed
to have a fascinating history. Even the writer and poet Robert Louis Stevenson
had made a visit there.

Alex advocated that the island needed recognition and how it should at least be
part of the programming conducted by the Fagatele Bay National Marine
Sanctuary. Aligning Swains Island with the even smaller sanctuary just made

sense to him. I think Alex's logic was that both of these small places seemed to share similar objectives, and that Fagatele Bay (being part of a federal government program) might have access to funds that could help protect, and perhaps even bring people back to, his beloved Swains Island. I remember thinking, at the time, that Swains Island was a distraction from how we might add value to Samoans and American Samoa, especially given the meager resources available. In addition, who would benefit from any investment in Swains Island? No one had even lived there since 2008. Alex, however, just kept pushing his case whenever he could, making a point to seek me out and grab my ear whenever I was in American Samoa.

Looking for the Real Samoa. Trying to understand how American Samoa actually worked was a gradual process for me, and I had many mentors along the way. The key had been establishing a genuine trust with the Samoans – a trust that we really had the same end goals as they did. Most importantly, it was demonstrating through actions that we actually "walked the talk." It is a ridiculously simple formula that makes sense anywhere. (I had learned this lesson decades earlier, when I lived and worked in the former Yugoslavia.) Applying the lesson in American Samoa, however, was different, and took longer to have an effect. It's a given that Samoan culture is completely different from Western cultures. At the same time, much of Samoan culture is hidden from view and remains very close to its origins. In addition, as affable as they are, Samoans do not necessarily "simply let you in" to see how they really view the world, which is through the lens of their traditional values and way of life, called *Fa'a Samoa.* Eventually, I was fortunate enough to be allowed to peek inside.

Over many years, the Samoans had learned how to deal with powerful Western cultures and their legacy of colonialism, while still standing apart and maintaining their fundamental Samoan center. If I had been born Samoan, I would like to think I would have been able to do the same. Knowing myself, however, I don't know. Surely, even today, the Samoans are wise to be suspicious of the motives of outsiders — just as they had been more than a century ago. They have a right to be covetous of their identity and culture on these small islands, essentially alone in the vast Pacific. The expeditions to Swains Island turned out to be a journey that revealed many aspects of Samoan history and its culture, both past and present.

Chapter 3 Pictures

Picture 3.1. Looking out onto tiny Fagatele Bay.

Picture 3.2. 2022 Sanctuary Advisory Council meeting in American Samoa at the Ocean Center. Note, retired Governor Togiola Tulafono at the head of the table serving as Council Chair.

Picture 3.3. The persistent Alex Jennings.

Modern American Samoa and Swains Island Rumblings

Alex was immediately excited about the prospect and began to lay out how he could arrange a "family" journey to Swains Island with Jean Michel Cousteau and Governor Togiola Tulafono. This would be a coup for Alex and the family. If only I could convince Jean Michel to come to American Samoa and then get him there. The building blocks for the First Expedition to Swains Island were beginning to take form!

Background. Although American Samoa is a very small place with a small population, hovering at around 55,000 people, it is deceivingly complex by Western standards. In comparison, independent Western Samoa (known as "Samoa" since 1997) has a population of nearly 200,000. Behind the facade of Western organization resides a traditional system of family, clan, village, and the leadership of *matai* (chiefs) that still determines how both Samoas work. It is a hierarchical society where birth and rank count.

The population of American Samoa is distributed among 10 major and 30 smaller villages, and nearly all land ownership is local. Most Americans would recognize or think they are familiar with how the island territory is governed. There is a territorial legislature, *Fono* in Samoan. It is comprised of 18 individuals in a senate, selected by Samoan customs and tradition across the territory, and 21 representatives in a house of representatives, each elected by district representatives from the 17 districts (*Picture 4.1*) comprised of villages. There is also a representative from Swains Island, which is considered equivalent to a district, although no one lives there now and it is more than 200 miles from Tutuila. Viewed from the outside, the consideration of Swains Island as a district might seem an anomaly. In 1982, however, when the Exclusive Economic Zones (EEZ) were established by the United Nations, extending the jurisdiction of sovereign maritime nations 200 miles seaward, Swains Island became a particularly important part of this U.S. territory. As part of the territory, the island considerably extends the EEZ and jurisdiction of American Samoa.

Traveling to American Samoa. American Samoa takes a while to get to, and, on average, there are only two flights a week from Hawaii. Miss a flight, either coming or going, and you're on an extended vacation. Then, getting to Swains Island is another matter entirely. There is no service of any kind to get there, and you need the family's permission to do so. Traveling to American Samoa takes planning. Traveling to Swains Island is nearly impossible.

Early on, I thought my travels to American Samoa were jinxed. Three times in a row I was airborne but never got there. The first time this happened was when I planned to attend the funeral of Governor Tauese Sunia in March 2003. One of our engines caught fire about an hour out of Los Angeles — and that was that. I had met with Governor Tauese Sunia only a few times, but his warmth, intellect, and infectious smile had begun to make me aware of how differently the Samoans thought about the world. I remember sitting in his office in Pago Pago and noticing how proud he was of a book about Samoa that he had almost completed. He passed away suddenly, and I was saddened to miss his funeral. He was only 62 at the time of his death. Tauese Sunia had also been a beloved and prestigious high orator, or *tulafale*, established during the Manu'a Kingdom era of Tui Manu'a. After two more failed tries, I finally returned to American Samoa and never again missed a date in Pago Pago.

On my first successful trip back, I got to meet the new governor of America Samoa, Togiola Tulafono, who was lieutenant governor under Tauese Sunia. Somehow I had never met him before (*Picture 4.2*). I didn't know it at the time, but my involvement and fortunes in American Samoa were about to change; the expeditions to Swains Island and so much more were to come. Togiola would remain governor for the next 10 years, and we became close friends and extraordinary colleagues. We remain good friends today, even across the great distances of the Pacific.

A Relationship Begins. Something had clicked between us. Everything else that followed, regarding how marine conservation and protected areas would add value to American Samoa and Samoans, came from this chemistry. Perhaps the most important outcome from our collaboration over those years was the 2012 expansion of the tiny Fagatele Bay Sanctuary (.25 square miles) to the National Marine Sanctuary of American Samoa, which encompassed 13,581 square miles of nearshore coral reef and offshore open ocean waters across the entire American Samoan archipelago. The expanded sanctuary also included six special protected areas, including waters around distant Swains Island. The courage and vision of my good friend Togiola were the force behind it all. The expeditions to Swains Island had helped to serve this greater purpose by drawing attention and interest to American Samoa.

In thinking of why Togiola and I clicked, I realize that being about the same age helped our chemistry. Our backgrounds and own journeys to meeting in Pago Pago that first day, however, could not have been more different. Togiola was born in a small village on the island of Aunu'u, off the southeastern corner of Tutuila, seven miles by water from Pago Pago. He had grown to manhood as a barefoot Samoan boy living fa'a Samoa by the sea. In those days, following World War II, most things in Samoa had not yet changed very much. I, on the other hand, was the son of an Italian immigrant family in Brooklyn, New York. Consequently, I am sure some individuals were puzzled by us and may have believed that we were "strange bedfellows."

One interesting coincidence we discovered was that our lives had probably first intersected at the 1964 World's Fair in New York City. Togiola was part of a troop of Samoan knife and fire dancers that performed at the Polynesian Pavilion at the fair for almost the entire year. It was Togiola's first experience abroad. I obtained my driver's license that summer and made numerous trips to the World's Fair with my friends, always stopping first at the Polynesian Pavilion to watch the action. As young men, my friends and I loved the knives and fire. In all likelihood, I had seen a young Togiola perform.

Togiola Tulafono. Togiola is about my height but perhaps a little shorter. As a Samoan, he is naturally stocky, but he always looks fit. The things about him that always stood out to me are his large, dark eyes and strong, powerful hands. If you look him in the eye, you will know what I mean. He is extraordinarily polite and retiring, as Samoan leaders tend to be. Nevertheless, you can almost feel the intensity in him. He has been in love with rugby his entire life, as a player, avid fan, and team sponsor. I would not have wanted to try to tackle him back in the day. When he came to the U.S. mainland to attend college, he brought his love of rugby with him. While teaching others about the sport, he also created a team.

By the time we met in 2003, Togiola had paid his dues many times over on his way to becoming Governor Tulafono. It is an interesting story of how a Samoan boy from a small village by the sea had gone to college, become a policeman in Hawaii, gone on to law school, become a lawyer, and then returned to Samoa to open a law practice. I think it was his unwavering determination that made him who he became, some of which I observed first-hand over the years. He also has always had an overriding goal from which he never veers: to do everything possible to add value to the lives of American Samoans. In my view, Togiola would always find a way to do the right thing, regardless of the political consequences. He never made anything about himself or sought any personal gain. It was always about fa'a Samoa.

Struggles for Sustainable Development. When we first began working together, I thought I was teaching him — and maybe I did a little. Later on, however, I came to realize that more often than not he had been the teacher and I the student. We spent much time together, not only in American Samoa, but in Hawaii, Washington, D.C., Florida, and other places. Our discussions inevitably explored ways to create sustainable development opportunities and programs for American Samoa, which had to be compatible with and respectful of fa'a Samoa. This meant a number of things. Most importantly, however, was that any money a project would generate had to stay primarily in or return to American Samoa. Also, any jobs that might come with a project had to further enhance the framework of Samoans' traditional family and cultural values. Lastly, there could be no detrimental effects on the terrestrial and marine environments of the islands. Developing sustainable projects to fit this model was difficult then, and still is today.

Given the remoteness and small size of American Samoa, many of the usual economic and sustainable development ideas or projects were just not realistic — and still aren't. Fishery canneries, with their ups and downs, had been the major industry and employer on Tutuila. In fact, many Western Samoans had come to American Samoa to work in the canneries that line the inner harbor at Pago Pago. Sustainable tourism, although hard to create, had always been considered an appropriate way to enhance American Samoa. Our discussions and this line of reasoning had ultimately led to the First Expedition to Swains Island.

Thinking about how Fagatele Bay could play a role in enhancing sustainable tourism in American Samoa really began to take form when, in 2006, Bill Kiene transitioned from science coordinator to superintendent at the Fagatele Bay Sanctuary. (Nancy Daschbach had left American Samoa after living there for 20 years.) Bill is a coral-reef biologist with a doctorate, but more importantly, at least to me, he is an experienced hand at working with indigenous communities in the Pacific and elsewhere. Bill pioneered many of the creative ideas that would eventually become a reality in American Samoa — from the creation, with the U.S. Park Service, of a coastal hiking trail for tourists (*Pictures 4.3,*

4.4) to the forerunner of the Sanctuary's Ocean Center in Pago Pago. Bill pushed ideas as far as a *Palagis* (a non-Samoan white person) could at that time. Not being a native Samoan nor speaking the language, and having no clan or village credentials, one could go only so far to enlist Samoans in new ideas. Bill had taken them pretty far. The barrier or intransigence that had helped the people maintain fa'a Samoa could also work against them at times. Nevertheless, I think it served them well overall.

Although The Papahanaumokuakea Marine National Monument had been created in the North Western Hawaii Islands by President George W. Bush in 2006 — after almost seven years of hard work by the Office of National Marine Sanctuaries and local citizens — and favorable eyes in government were beginning to look towards the Pacific, there was no discussion of expanding this tiny place or doing anything in American Samoa.

A Catalyst Arrives. Working with Togiola (that is, Governor Tulafono) and the Samoans in his cabinet, a very experienced Samoan woman and coastal zone planner with the territorial government was encouraged to compete for the job as Bill's replacement. After a lengthy and open competition, Gene Brighouse (*Picture 4.5*) had risen to the top. In 2008, she became the superintendent of the Fagatele National Marine Sanctuary, following in the footsteps of Nancy Daschbach and Bill Kiene. I knew Gene only vaguely at that time, but her credentials were impressive. She had both undergraduate and graduate degrees and was managing the entire Coastal Zone Management Program in American Samoa. Gene had grown up in Apia in what was still known then as Western Samoa. Like so many Samoan families, hers was spread across both American and Western Samoa.

Besides the relationship that Togiola and I had formed, the serendipitous event of meeting Gene led to my next most important partnership in American Samoa. I think Togiola may have known it would work out that way. Gene and I also clicked, and I eventually gave her the nickname "Big Sister," because of the way she seemed to care for everyone in her fa'a Samoa manner. She also knew the territorial government inside and out and had personal experiences with the people and leaders in all the coastal villages. Later on, I would often describe her as the "missing piece of the puzzle." When all was said and done, Gene became the catalyst that made everything work, including the expeditions to Swains Island. I was fortunate that she had decided to join us. Like Togiola, she never made anything about herself or sought any personal gain. I guess I have always liked people like that. I still affectionately refer to Gene as "Big Sister" today.

It was in this atmosphere that we really began to mull over and experiment with ideas of sustainable development in American Samoa. It was also when Alex Jennings again began to present ideas about Swains Island. He had raised the proposition that Swains Island could become a special destination for off-the-grid sustainable tourism experiences. Gene encouraged me to consider if there was some way we might help Alex. At the time, I don't think Alex had thought the idea through. The remote and exotic South Pacific nature of Swains Island, however, could be worth investigating on this score, but we had precious little information about Swains Island beyond what Alex and others had told us. The first step was to find out as much as possible about the island.

Finding Out More about Swains Island. Hans Van Tilburg is a marine archeologist I was fortunate to recruit to the Marine Archeology Program evolving within the National Marine Sanctuary System (*Picture 4.6*). Hans lives on Oahu in Hawaii and is a highly regarded marine archeologist across the entire Pacific. I have known many archeologists from around the world, but Hans stands out as if cut from a different cloth. I think it comes down to his demeanor, patience, and quiet confidence. Perhaps it emanates from his background or because he came late to the profession. Whatever it is that makes him different, he is a person you don't forget. Unlike many others, he is rarely selling; and, if he does, it's a soft sell. Hans is about six feet tall and his dark, slightly slanted eyes hint at his Chinese heritage. He is solid of build, an excellent diver, and a good hand in the field. In fact, Hans and I would end up diving together many times, including Swains Island. But this was 2008, and all of that was yet to come five years later on the Second Expedition to Swains Island.

I had the good sense to assign Hans to conduct a background study of Swains Island. He was to find out everything he could and put it all together in a report. The goal was not only to give us a basis for further considering Swains Island, but also to provide Alex and his family with an objective document about their island. I left it to Hans to work it out; he was more than capable. Hans made his first trip to Samoa and met with Alex and other members of the Jennings family to gather what he could from family history to help guide him in his research. Early in 2009, Hans completed the background study, which opened a door on many aspects of the island's past — some of it unknown even to the Jennings family. Knowing more about Swains Island was one thing, but who might provide insights about sustainable tourism in such places in the far South Pacific? This is where my friend and colleague Jean Michel Cousteau joined the story (*Picture 4.7*).

Enter Jean Michel Cousteau. Jean Michel is an icon in ocean conservation. Sometimes I think people still confuse him with his more famous father, Jacques Cousteau. Jean Michel worked with his father on most of his adventures and played an important role, mostly behind the scenes, in pulling them off. After his father's passing, Jean Michel continued the quest as a filmmaker, educator, and ocean advocate. We have worked closely together many times, even though I was in the government. He's always supported me if he could. Jean Michel is generally not a fan of government bureaucracies — but then, who is? What most people don't know about Jean Michel is that he is also a classically trained graduate architect. Architecture and urban planning were always his passions. Nevertheless, he dedicated himself to the family business when his father asked him to do so. One of his notable accomplishments is that he has a unique eye for sustainable tourism, especially in the Pacific where he helped create sustainable tourism in Fiji. He is an advocate of totally "green" economics. And although he became an American citizen, he remains French at heart.

Enlisting the Governor. Alex and his Jennings family members were excited to get Hans's report. I think it was validating to them. Before I mentioned to Alex that I was thinking about enlisting Jean Michel to the Swains Island cause, I decided to speak with Togiola to get his views. It was a good thing I did. Togiola was, of course, quite knowledgeable about Swains Island, and had represented the Jennings family as a lawyer. He also had been to the island once. Togiola had reviewed many of the past schemes proposed for Swains Island and said, a little wistfully, that he would like to go back someday and even scuba dive there. No governor of American Samoa had ever been a scuba diver, and no one was known to have dived at Swains Island up to that time.

As it turned out, I became instrumental in getting my friend Togiola into scuba training with a U.S. Park Service employee and certified instructor living on Tutuila. There were no recreational diving operations in American Samoa. I never had enough time during my visits to fully train and certify Togiola myself, but we did get in the water together a few times. Togiola was incredibly knowledgeable about all things in the terrestrial and marine ecosystems of Samoa. I think he became a keen observer and student of the natural world at a very young age in his village by the sea. In his youth, he was an accomplished free diver on the reefs around Samoa. He taught me a little free-diving trick about how to store air in your cheeks for an extra breath. I tried it, and it worked.

Togiola was not a scuba diver and as governor he could hardly get the time to practice. He knew his limitations and decided the only person he felt safe to dive with was me. Diving around Samoa could be a challenge for men half his age. I was flattered. By that time, I had been instructing and teaching all aspects of scuba diving for more than 20 years. He knew I was absolutely no nonsense — governor, friend, or not — and so did his wife, Mary. Mary also insisted that her husband was not allowed to dive with anyone else. His diving frightened her. I promised her that nothing would ever happen to Togiola on my watch.

When Alex and I sat down again, I mentioned how I thought I might get my friend, Jean Michel, involved. Alex beamed and immediately suggested that Jean Michel visit Swains Island. He was nothing if not a big thinker, especially about his beloved island. (He and his older brother, David, had lived there as children, and Alex lived there even longer.) Alex was immediately excited about the prospect and began to lay out how he could arrange a "family journey" to Swains Island with Jean Michel and the governor. This would be a coup for Alex and the family. If only I could convince Jean Michel to come to American Samoa and then get him there. The building blocks behind the First Expedition to Swains Island began to take form.

Chapter 4 Pictures

Picture 4.1. Map showing the complexity of political districts in American Samoa.

Picture 4.2. Governor Togiola Tulafono on Swains Island.

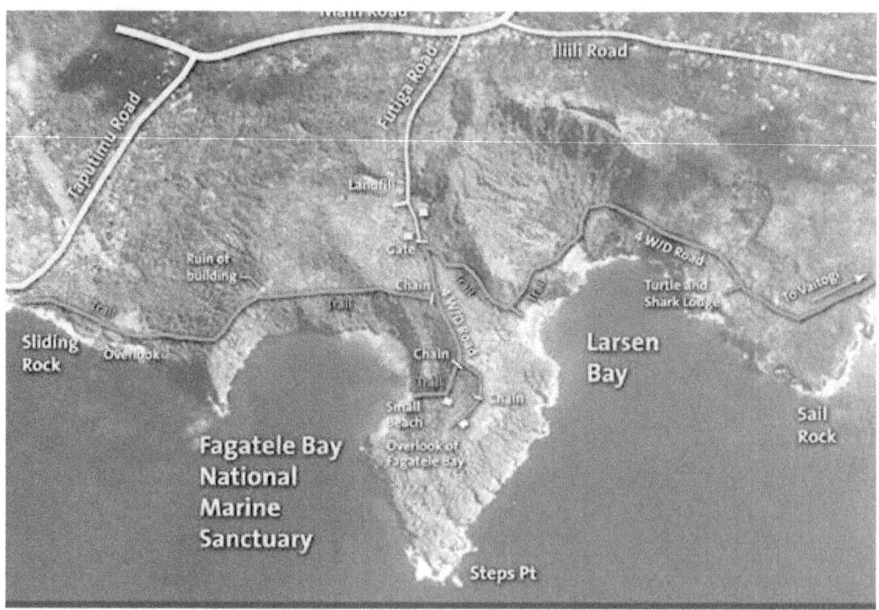

Picture 4.3. Fagatele Bay trail pioneered by Bill Kiene.

Picture 4.4. *From left:* The governor, author, and Bill Kiene opening the trail. The brooms are used for ceremonial sweeping of the trail.

Picture 4.5. Gene Brighouse in her office in Pago Pago.

Picture 4.6. Hans Van Tilburg, looking the part of the archeologist.

Picture 4.7. Jean Michel Cousteau is intrigued to hear of Swains Island.

Chapter 5

The First Expedition Begins

The morning after my arrival in Pago Pago, the loading of the Lady Naomi was well underway. I think the expedition was all the buzz on Tutuila. It was bittersweet because Jean Michel Cousteau was not on the dock that day.

Background. Throughout all his travels in the Pacific, Jean Michel Cousteau had never heard of tiny Swains Island — a remote, uninhabited speck more than 200 miles north of Pago Pago in the U.S. Territory of American Samoa. As I explained to him the island's remoteness, a brief history, and then the ideas Alex had expressed, Jean Michel became interested enough to look at a map and ask me some questions. After reading Hans's report on Swains Island, Jean Michel said, "I am intrigued. Why haven't I heard of this place?"

The invitation that followed to visit the island, investigate the potential for a totally green, off-the-grid, sustainable tourism operation, and to accompany the governor of American Samoa there, clinched it for Jean Michel. Yes, indeed, he was in for an expedition to Swains Island. He wanted to explore it and, most importantly, dive its remote reef to see what secrets the corals might reveal to him. The governor's presence on the expedition had also given credibility to the idea. Maybe something might actually come of this? The wheels began to turn about how such an expedition could be accomplished.

It was early 2010, following President George W. Bush's 2009 proclamation that the government would work to create and expand the tiny Fagatele Bay National Marine Sanctuary. It had taken a lot of hard work to get to this point, and it would continue to take a lot more work by many. Presidential proclamations don't just happen. But our efforts were starting to pay off and eyes were now on American Samoa. An expedition to Swains Island was news.

Planning Begins in Earnest. Three groups came together to organize the expedition to Swains Island. First and foremost were Alex and the Jennings family. They began planning for a major family visit to their home island. They had experience with such family outings, in part because they received an annual stipend from the territory of American Samoa to periodically visit the island. The visit we were now planning was going to be different, though, as it was more complex and involved more people than usual.

The ways of the Samoan family and practice are very different from those of Westerners. There is a certain communal sharing within a family that, for example, allows a family member to move into another member's home seamlessly, with little said. Individuals within a family have a tradition of communal rights. Consequently, a larger-than-usual family contingent would make the trip to Swains Island, and other family friends would also be invited. It would be a joyous event for many family members to reconnect or to be introduced to their home island and their past.

There were many logistics to organize for this journey, including finding a ship. The ship usually used for the family visit, the vessel M/V *Sili*, an American Samoa government vessel, was out of commission for major repairs. Then there was everything else to acquire, organize, and pack to bring to the island: food, fuel, generators, tents, bedding, and much more. I would not describe the gathering and organizing as being conducted with anything like "military precision," but rather the whole affair was more like an elaborate family picnic or reunion to which everyone just "chipped in." It was the Samoan way — and it worked.

The second group was the governor's contingent, which turned out to be considerable. I had no idea how the governor's presence would affect the size and conduct of the expedition. The effort to pull this group together was overseen by Lelei Peau, who had become the governor's right-hand man, working closely with him on all coastal matters in American Samoa and on many other things (*Picture 5.1*). Lelei had also been Gene's boss when she was at the Department of Commerce in the territorial government. Like Gene, Lelei had proven to be a catalytic force in many things, and they always worked well together.

Lelei was a strict protector of the Samoan way and certainly had his own ideas. He taught me a lot about Samoan protocols. He was university-trained at both UCLA and the University of Hawaii. On many occasions, he was asked by the governor to represent or to speak for American Samoa. He always seemed to be nearby, if he could be, when Togiola and I were together. A moderately sized man for a Samoan, he was about six feet tall, bald-headed, and stocky; and he had the broad, round features of many Samoans. Over the years, I got to know him pretty well. Some years later, I would actually hire Lelei into the federal service. Later, he was elevated within the Samoan *matai* (chief) system and given a new name: Atuatasi. He is now known as Atuatasi-Lelei-Peau. Both these things would take place much later. Thus, I refer to him as Lelei during this period.

An entire contingent of the Samoan marine patrol and its equipment, including two Jet Skis and a small launch, would also accompany the governor on the expedition. The governor's chief of security, Lelei, an emergency medical doctor on a three-month assignment in Pago Pago, and a few others rounded out the group. I don't think the others had a function other than just being there. It seemed a big deal for everyone that the governor was to accompany the expedition. Not only did he represent the highest official authority in American

Samoa, he was also a high-ranking Samoan high chief for his family, village, and country. In Samoa, this meant that a series of other, lesser chiefs or village heads had elected him as their leader, in accordance with their customary laws and traditions. While this was certainly of great importance to the Samoans, it was of little consequence to most Westerners.

The third group was the Fagatele Bay National Marine Sanctuary staff, now managed by Gene, who was responsible for organizing logistics and support for all of the diving operations. Given the governor's intent to dive at Swains, I personally oversaw this part of the planning. This may have seemed odd to a few people, but I had more experience adventuring and diving across the Pacific than anyone — except perhaps for Jean Michel — who would be making the trip. I would take no chances on the dive operations, and would personally lead the governor's underwater adventure. Yes, he was the governor, but more importantly, Togiola was my friend.

In Samoa, Kevin Grant, who was on Gene's staff, coordinated the scuba gear and tanks. He also arranged for a National Park Service EMT, who was stationed at the National Park on Tutuila, to accompany us. The EMT was also a certified diver. I rounded out the dive team with Paul Chetirkin, an underwater videographer and strong diver from our West Coast National Marine Sanctuaries. Paul would document our dives with the governor. I had no worries about Jean Michel. We had dived together before. Besides, he was — and still is — the most experienced and longest-diving human on the planet. I don't think anyone will ever break his record, and he is still diving. Gene played a central role and coordinated all elements of each group with Alex and the Jennings family, Lelei, and the governor. A few other members of her staff would also make the once-in-a-lifetime journey to Swains Island. Always the good soldier, Gene would not make the trip; there was still work to be done in Pago Pago.

Jean Michel Is Forced to Abort. One week before the expedition was scheduled to leave Pago Pago, Jean Michel had a family medical emergency. Unfortunately, he couldn't leave the West Coast. It was a hard choice for him, but he made the right choice. The expedition was poised to launch, and we would go without him. At the time, it seemed he might never get an opportunity to visit this faraway place. (Little did any of us know then that a Second Expedition would be in the offing.)

Alex Jennings, with (I suspect) a little help from the governor's office, had leased an outstanding vessel for the expedition. An inter-island ferry from the Samoan Shipping Corporation, the M/V *Lady Naomi* (*Picture 5.2*) is a large vessel with a cavernous cargo hold for vehicles. She has a sleeping compartment for 35 overnight passengers, an excellent galley and dining room, and viewing compartments and benches throughout. With the acquisition and scheduling of the *Lady Naomi*, our preparation for the First Expedition to Swains Island swung into full speed. It was now "for real," and there was no turning back.

At the Dock. The morning after my arrival in Pago Pago, the loading out of the *Lady Naomi* was well underway (*Picture 5.3*), and the expedition was all the buzz on Tutuila. It was bittersweet because Jean Michel was not on the dock that day. Several days before, his longtime partner, Nan, had been rushed to the hospital for a medical emergency. Jean Michel had rushed to her side and could not make the flights to journey out to American Samoa. There are only two flights a week from Hawaii to Pago Pago and no other options. In addition, the *Lady Naomi* had to sail on schedule. Jean Michel's participation helped provide the stimulus for the expedition, and now we would sail without him. The call of adventure to Swains Island, however, would continue to draw him to that remote island, and to that very same dock.

The number of people and amount of equipment going aboard were more than I expected. The ship had backed into the dock and the rear tailgate on the motor deck was raised (*Picture 5.4*). Members of the marine patrol (*Picture 5.5*) loaded their new Jet Skis and equipment aboard. Jennings family members and others slowly filed aboard. Well-wishers and family members filled the dock (*Picture 5.6*). Gene and her team were there, ensuring that all ran smoothly. Kevin Grant, Paul Chetirkin, and a few others carefully loaded and stowed the scuba equipment and tanks in the cargo bay. Being among the last to board the *Lady Naomi*, I had just been taking it all in (*Picture 5.7*). The scene was unlike that of any other expedition I had been on.

Just before going aboard, Alex reintroduced me to his older brother, David Jennings (*Figure 5.8*). David is former U.S. Navy and lives in Texas. He had probably arrived the night before on the same flight from Honolulu that I had been on. It was the first of many meetings between us. Two years later, David would play a key role in organizing and participating in the Second Expedition.

The *Lady Naomi* wasn't a new ship, but neither was she very old. Once on board, it was first things first: find a bunk in the berthing compartment. The compartment has two corridors with bunks on either side and curtains facing the corridors. The compartment and bunks were clean and there was storage above and below. I remember it had been warm in the compartment. Later, during the night, it actually got a little chilly: the *Lady Naomi* was air conditioned. Although most of the bunks were already occupied by the time Paul Chetirkin and I got there, we found two bunks on the bulkhead side of the second corridor.

Kids, grandmothers, young moms and dads, and others were all settling in. It was, indeed, a family affair. They also brought lots of stuff in bags and baskets, which they squeezed in here and there. Everyone was friendly, all smiles, and talkative. Fortunately, all Paul and I brought aboard were light backpacks. Of course, Paul had his camera and lights, which he also stored in his bunk. The governor, Togiola, slept elsewhere. The ship's captain, Captain Wally, had graciously given him his cabin. I think Paul and I both enjoyed being in the berthing compartment along with everyone else.

As the *Lady Naomi* left the dock and glided through the calm waters of the harbor to the open ocean, everyone was on deck to watch the lush green mountains surrounding Pago Pago slip by (*Picture 5.9*). It was a lazy early afternoon in American Samoa. Alex walked with me all over the ship to be sure I met everyone. Togiola and I then found a spot, forward on the upper level, to sit and watch the blue Pacific as the *Lady Naomi* steamed along *(Picture 5.10)*. Once out of sight of Tutuila, we saw nothing but blue sky and sea for the next 200-plus miles. It would take almost 24 hours to get to Swains Island. We had purposely departed around mid-morning to arrive at the island about the same time the next day.

It was a calm sea. When the sun set, a brilliant starlit night emerged that can be seen only far out at sea. The stars seemed to come right down to the sea and surround the ship. It was hard to leave the deck that night and go below to our berthing compartments. Some of the Samoans slept on deck, including all of the marine patrolmen. The contrast between this pleasant voyage to Swains Island and the Second Expedition, to come two years later, could not have been more extreme.

Early the next morning, the day broke bright, sunny, and warm. I am sure this contributed to the festive mood aboard the *Lady Naomi* as we approached Swains Island. Most influential was the Jennings family's joyful attitude as they returned to their home island, if only for a short visit. For many of them, it was their first visit back. For the young ones, it was their inaugural visit to their legendary home island. It was both a celebration and a validation of themselves. Their smiles were infectious. Then, suddenly, everyone moved forward on deck and strained their eyes to see tiny specks beginning to emerge on the horizon *(Picture 5.11)*. The specks were the tops of coconut trees emerging from the sea — on Swains Island.

Chapter 5 Pictures

Picture 5.1. Lelei Peau, a catalytic force.

Picture 5.2. The *Lady Naomi* at the dock.

Picture 5.3. The car ramp made for easy loading and transferring of people and equipment into the skiff for the trip ashore.

Picture 5.4. Loading the *Lady Naomi*.

Picture 5.5. Two members of the marine patrol and their Jet Ski.

Picture 5.6. Gene Brighouse, *left front*, and Lelei, *right*, on the dock before Lelei boards the *Lady Naomi* for the First Expedition.

Picture 5.7. The author, *left*, and Kevin Grant watch the loading process.

Picture 5.8. David Jennings is all smiles on the dock. He will be a key leader on the Second Expedition.

Picture 5.9. Watching the green hills go by. Captain Wally, *left*, sits wearing sunglasses. Governor Tulafono, *center*, leans against the bulkhead.

Picture 5.10. Leaving Pago Pago behind.

Picture 5.11. Swains Island emerges.

Chapter 6

First Footsteps on Swains

Swains Island is not a place where you can just walk into the water and dive. There's the reef, strong currents, and it immediately gets very deep. The nearest medical facilities are 200 miles away in Pago Pago. The closest recompression chamber is in New Zealand, and the Lady Naomi is the only way out. The island is also beyond the range of helicopters. If something goes wrong out there, you are essentially on your own — and a situation can go very bad quickly.

No sooner had the tops of the coconut trees on Swains Island been sighted than preparations for landing began. Everyone on the ship was in motion. The berthing compartment was crowded and chaotic as people packed up their things and looked for family members and companions. Everyone and everything moved through the ship to the motor deck in anticipation of our arrival.

A very narrow reef surrounds the island. It is so tight to the shore that there is little to no room between it and the island (*Maps 5-8*). Outside of the narrow reef, the bottom quickly drops steeply to great depth. The *Lady Naomi* and most other ships cannot anchor at Swains Island so the ship remained at sea, steaming close by, though often out of sight over the horizon. We landed on the leeward (downwind) side of the island, where a single aluminum skiff could maneuver through a narrow cut in the reef back and forth from the *Lady Naomi*. This was the side of the island where people had always landed and *copra* (dried coconut kernels) was painstakingly taken out over the beach to ships standing by. The gradient of the beach here is very gentle and comprised mostly of hard coral debris. A little sand berm rises up to meet the coconut trees (*Picture 6.1*). The island's village (known as Taulaga), copra works, and other buildings were located close to this beach. The village church still stands. The remains of the Jennings family home (or "Residence," as it is called), located on the other side of the island, is connected to the beach by a path (once a road) that ran along a narrow corridor between the lagoon and the beach.

Getting Ashore. On the voyage out from Pago Pago, I learned from Alex that a small advance party of family members had been on the island for a couple of weeks, preparing for our visit. Apparently, there had been a lot of cleanup to do. The aluminum skiff was used to bring and unload everyone and everything onto the island. The skiff looked a little battered or maybe just well used. A grizzled family member, who was expert at navigating the cut through the reef, brought the skiff out to the ship (*Pictures 6.2, 6.3*). The "cut" wasn't a channel or straight shot — just a route between coral heads. The skiff was the largest boat that could make it through safely. Then there was the tide to consider as well. At low tide, even the skiff would scrape bottom.

The *Lady Naomi* loitered as nearshore as she could (*Picture 6.4*). The rear tailgate on the motor deck was raised, and people and supplies were lined up to get into the skiff and make the trip back through the cut to the beach. The marine policemen were the first off with their new Jet Skis. They were excited to get going and followed the skiff in through the cut (*Picture 6.5*). They proceeded to make trips back-and-forth to the *Lady Naomi* to help get people ashore. Life jackets were issued to everyone. It took quite a while to complete off-loading with the single skiff, and it was near mid-morning when the *Lady*

Naomi departed to cruise lazily nearby *(Picture 6.6)*. Occasionally, she would pop up on the horizon, disappear, then pop up again somewhere else.

The Work Was Just Beginning. Once ashore, our work really began. First, everything had to be carried across and up the beach, maybe 50 to 75 yards, to the sand berm and under the coconut trees. Then, everything had to be carried to places further inland, where the old village had once been or elsewhere. A building or two remained that were kept up and still useable. One long, open building had screens and appeared to be used as a sort of family dormitory for sleeping. Another smaller building, also open, was used as a kitchen and had a small portable generator nearby.

Many things, such as the dive equipment and tanks, were stored under tarps a little inland from the berm. A line of tents and camping equipment was set up along the top of the sand berm under the shade of the coconut trees *(Picture 6.7)*. This is where we stayed. If one walked 25 yards further in from the tent line, the buzzing and biting bugs were thick. The breeze off the ocean was just enough to keep them back — most of the time.

A lot of work had been done by the small advance party, but much still remained. The younger boys who arrived with us were all assigned jobs. Thousands of rotten coconuts and fronds littered the ground. The advance party had removed all they could from the immediate area and placed them in large piles further up the beach to burn, partly to chase away mosquitoes. This process continued, and the marine patrol also pitched in. The piles were at least 10 feet high. Other family members had hiked across the island to the original Jennings Residence and cemetery to clear away the encroaching jungle tangles. Holding back the jungle is a continuous labor in the South Pacific, even on tiny Swains. Then there was the food preparation work required to feed everyone. Something was always being prepared throughout the day. I briefly walked around with Alex to get the lay of the land and take a peek at the enclosed "freshwater" lagoon. Since the plan was to conduct the governor's dive sometime around mid-afternoon, however, I had to leave exploring for later and get busy.

I didn't believe anyone had ever scuba dived the waters off Swains Island before our arrival. Certainly, no one I knew had done it. No one in American Samoa or on the expedition had heard about anyone who had, either. Later, however, a NOAA research ship planted a small underwater plate somewhere on the reef at Swains Island. The plate was to attract marine growth as part of a Pacific Ocean monitoring program. This had to have involved a scuba dive or two. Jean Michel Cousteau and I would make a point of finding and documenting the monitoring plate two years later on the Second Expedition.

Setting up to Dive the Reef. Swains Island is not a place where you just walk into the water and dive. There's the reef itself to contend with, strong currents, and it immediately gets very deep. As mentioned, the nearest medical facilities are 200 miles away in Pago Pago. The closest recompression chamber is in New Zealand, and the *Lady Naomi* is the only way out. Swains Island is also beyond the range of any helicopters. If something goes wrong on Swains, you are essentially on your own, and a situation can go very bad quickly. In fact, in order for me to be on this trip and dive I had to take annual leave. Because I was an active-status NOAA/U.S. Government diver, had I been on duty, I would not have been permitted to dive under such circumstances. It's a good rule.

By this time, however, I had been on or conducted many nongovernmental expeditions throughout the Pacific and felt confident I could ensure our safety. I learned through experience to plan these dives very carefully, to always dive the plan, to be prepared for all contingencies, and, most importantly, to not make mistakes. My friend Togiola, the governor, would be safe on my watch. I had made it clear to everyone before we left Pago Pago that it was my way or no way. Writing now, it seems heavy-handed of me, but this was serious business. No one on the expedition seemed to have a problem with my position — at least not that I heard.

First, I had to determine where we would dive, and then how we could dive it safely. Surveying the reef from the beach was not enough. No one had a clue about the topography and currents outside the reef. While others were still organizing the camp, I grabbed Paul Chetirkin to do a reconnaissance with me along the reef face, outside of the cut, and along the coast. We assigned Kevin Grant to keep an eye on us from the beach with binoculars.

With masks, snorkels, and fins, we swam our way through the cut and then along the reef edge. The visibility was beyond great — we could clearly see 150 feet or more. We swam along the steep drop-off and in and out of small canyons, free-diving here and there. Paul and I periodically lingered on the surface to talk over the conditions and get our bearings with features on the shore. Eventually, we decided on a good route for the governor's dive and discussed how Paul would take video of the governor. We then swam back along the reef, through the cut, and to the beach. It was clear that these could be challenging waters, even for good swimmers.

When we returned to our camp under the coconut trees, everyone had gathered round, including Togiola. My first order of business was to make it clear that no one was permitted to swim or snorkel outside the reef without permission, that they must be accompanied either by the skiff or a Jet Ski, and that someone on shore had to observe them. A swimmer could easily be carried away off the reef at Swains Island — but, again, not on my watch. Togiola nodded his approval. Nevertheless, I could see in the eyes of some that look of: "Who is this guy to cramp their style?" I hadn't asked them what they thought nor how well they could swim, because it didn't matter: there could be no foolish mistakes out here. I don't know of anyone who violated the rules.

Diving with Military Precision. Finally, it was time for the governor to go diving. I thought the swim out and back, especially with scuba gear, would likely be difficult for Togiola, especially if the longshore current flipped. Consequently, I laid out a step-by-step, "almost-military" plan. First, there would be only four divers: Togiola, Paul Chetirkin, the park service EMT, and me (*Picture 6.8.*). Second, the skiff would be loaded just off the beach with all of our gear and tanks. The skiff would be navigated to the entry point by the most experienced boatmen on the island. Third, two Jet Skis driven by the marine patrol would accompany us, and one of the Jet Skis would tow a rescue platform. Fourth, the four divers and boatmen would ride the skiff, accompanied by the Jet Skis, through the cut and along the coast to our entry point.

Once at the entry point, the EMT would be first into the water, then Paul, then Togiola, and then me. I would stay in the skiff until everyone was in to serve as a rescue swimmer, if needed. We would put on our scuba gear in the water to avoid the possibility of capsizing the small skiff. Once Paul and the EMT had helped Togiola with his gear and given me the "OK" sign, I would roll in and gear-up. The skiff and Jet Skis would then follow us by watching our bubbles and try to remain close, but not directly over us. Some people in the entourage, such as Lelei, wanted to dive with the governor or accompany us in the skiff, but I simply said no. Kevin Grant would be our spy on shore. No mistakes! Lastly, after briefing everyone, which included drawing a diagram of the route and positions (especially for the skiff and Jet Ski drivers) and making sure we all drank plenty of fluids, we were ready to go. Togiola grinned from ear to ear as our scuba gear was carried down to the skiff by well-wishers and marine patrolmen.

We pushed the skiff along to get into deeper water so we could lower the outboard motor, and soon we were all aboard. The Jet Skis followed as we carefully maneuvered through the cut. It was a good day to dive: sunny, no wind, and the water was warm. At the appropriate spot, we lined up with an outcropping on the beach that Paul and I had used earlier, cut the motor, and got into the water. All went as planned. Before we left the surface, I exchanged signals with the marine patrol to make sure the Jet Ski contingent knew what to do. Of course, they did; these Samoans were honored to be along to help protect their governor. Everyone showed a palpable and touching reverence to Governor Tulafono.

Once in the water, I guided Togiola down to about 30 feet and began our swim along the steep-walled edges of the reef. Conditions were spectacular. The EMT followed close behind us. Paul swam below and then out to the seaward side and back as he positioned himself to shoot video. Having trained many divers in similar "wall" environments, I knew to keep the student between me and the wall. When Togiola veered seaward, I gently guided him back toward the wall, often dropping back or moving up to get out of Paul's video frame. I think this dive was a once-in-a-lifetime experience for Togiola. He was enraptured by all that he saw and recognized. Once or twice I gently signaled him "upward" as he descended into a deep pocket, investigating something of interest. A shark came in and got his attention. He wanted to get closer and follow it. Togiola's large, dark eyes looked even bigger in his mask, and they seemed to be aglow. It was not fear, which I had seen many times before, but rather, rapture. After about 30 minutes, Paul signaled that the video was good. Togiola then signaled me that he had about 1,000 psi of air remaining. It was time to call the dive. We made our ascent to a safety stop, and then returned to the surface.

Togiola looked good as we surfaced and he bobbed in the gentle waves, but I decided it was prudent not to get him out of his scuba gear. The EMT and I helped him up onto the rescue platform being towed by one of the Jet Skis. It takes a little more technique to power-up and onto a skiff. Keeping his gear on him also ensured he would be attached to a buoyancy device and would float if anything unforeseen happened. The platform was low in the water and easy to climb onto. Togiola was all smiles as he relaxed on the platform and we three got back into the skiff. It seemed almost like a triumphal procession, motoring along the reef and back through the cut. Several people waited onshore, and as soon as the Jet Ski towing Togiola got into shallow water, they swarmed around him and removed his gear. My friend stood there in the shallow water and waited until we all got in. He then shook everyone's hand, thanking them. We walked together up the beach to the shade of the coconut trees. Mission accomplished.

Right then, I was feeling pretty good about things. Togiola was a novice diver, and he had done well. I had fulfilled my promise to his wife, Mary, and he was safe and sound. Paul went on to make an excellent video from the footage he gathered on the dive. The video of the governor of American Samoa diving at remote Swains Island showed in the welcome center at the airport in Pago Pago for many years. It is still showing in the Ocean Center in Pago Pago. Even today, very few people can claim to have dived Swains Island — and, to my knowledge, no other Samoan governors or paramount chiefs have done so.

Chapter 6 Pictures

Picture 6.1. The gentle slope up to the coconut trees.

Picture 6.2. Pulling the skiff to shore.

Picture 6.3. Pulling the skiff ashore. Note the *Lady Naomi,* in background, and the cut through the reef, *left,* where a Jet Ski has just come in.

Picture 6.4. The *Lady Naomi* comes as close to shore as she dares.

Picture 6.5. Lelei wades ashore from a Jet Ski.

Picture 6.6. We have landed. *From left*: Alex Jennings, Governor Togiola Tulafono, and the author.

Picture 6.7. Tents under the coconut trees where we stayed.

Picture 6.8. The dive team on the *Lady Naomi. From left:* EMT, Governor Togiola Tulafono, Kevin Grant, the author, and Paul Chetirkin.

Chapter 7

Exploring the Island

Then Captain Wally dropped a bomb that changed everything and put me on the path that led to a Second Expedition. He said: "My grandmother she told me her mother told her that there had been a channel into the lagoon. Yeah, a channel! But a big hurricane — maybe hundreds of years ago — come and close it." While Captain Wally talked, he pointed toward the shallow end of the lagoon, across the island where the day before we had seen the tupua. It was as if the Earth had stood still for a moment, and everything I had seen and heard on the island came rushing together in my mind.

Background. I had known something about Swains Island before we ever set foot on it. Alex and others had told me a great deal about the island over the few years I had known them. In addition, Hans's report provided an excellent background, even for the Jennings family. Walking the ground was a totally different matter, though. It stimulated all of the senses and revealed things that no one had spoken about, didn't think were important enough to pass on, or perhaps had not even noticed. It was walking the ground on that very hot afternoon that revealed pieces of the Swains Island story and insights that got me thinking that artifacts from the island's past might lie hidden in the lagoon.

We Go Exploring. It was late afternoon when Alex, the governor, Lelei, I, and a few others began to explore the island *(Maps 7 and 8).* Alex, of course, led the tour. We began in the area of Taulaga village *(Picture 7.1),* which was directly behind the line of tents on the landing beach. Swains Island must have been a going concern during the copra era and probably even before. All was now gone except for foundations, a few chimneys, and imprints on the ground. A large open area, however, still remained. Obviously, the advance team had been busy cleaning up more than just coconuts. At the north end of the open area had been Taulaga village. The copra drying sheds, furnaces, and other working buildings all seemed to have been located in the middle of the open area. The church, still standing and intact, was on the south end *(Picture 7.2).* Lots of rusting items poked out of the jungle on the fringes of the open area; and, for that matter, they could be seen everywhere around the island. Most of these were primarily remains from the plantation period, and some from the earlier whaling period. The water tank, built in the 1920s to provide fresh water for the village, still worked and served as an important water source for visitors to the island *(Picture 7.3).* It was a lot to absorb on a casual stroll.

Toward the back of the open area, a cleared path led to the enclosed freshwater lagoon. I could see why some called the lagoon a lake *(Picture 7.4)* — it was fringed by trees and entirely open, except for a very small manmade island near the shore opposite of the village. The lagoon was a little over one square kilometer in area, but seemed larger from the shore. There are not many enclosed freshwater lagoons on islands in the Pacific. Obviously, the lagoon had played an important role in the lives of those who had lived and worked on the island for generations. At this point, however, not a lot appeared to be remembered in any detail.

Poles or pilings that had supported drying racks for clothes and shells still stuck out of the water near shore, where the path met the lagoon *(Picture 7.4).* Swains Island shell necklaces are still prized in American Samoa. What initially stuck in my mind, though, was that locals said that if you threw a coconut log into the lagoon, it would become "petrified." The science party on the Second Expedition would reveal so much more about the lagoon and the island itself, but that comes later in the story.

My education really started when we began the walk on the old road (*Picture 7.5*) around the lagoon, partially cleared by the advance party, to the other end of the island to the old Jennings family residence. This area is called *Etena*, which is Samoan for "Eden." It was more like a path than a road, in many places. Alex called it the "ring road." The path went into the heavily jungled area to the south between the lagoon and the shore. It was incredibly hot and insect-heavy as we walked along. Sweat poured off of everyone. (In hindsight, I think we should have brought extra water bottles.) Soon after getting on the ring road, Alex pointed to the graveyard for both the Solomon and Tokelau islanders who had lived and worked on the island. At the time, both burial places were covered with overgrowth. The Jennings family cemetery (*Picture 7.6*), which was cleared, was further down the road near the old homestead.

As we walked along, Togiola seemed to know every tree and plant and their uses. Besides being governor, he was a Samoan naturalist. The jungle on this side of the island appeared to be indigenous growth. It seemed that this small area had never been converted into coconut palm trees during the plantation period. Togiola pointed out the abundant native hardwoods necessary for building and repairing Polynesian homes and boats. He also pointed out wild taro and tapioca plants (traditional food sources) growing naturally off the road. Whether these plants were actually indigenous to Swains Island or brought by early Polynesians or others is unknown. We also picked and sampled a wild lettuce as we walked. At one juncture, Alex pulled back some undergrowth and showed us a shallow well just off the road. It was maybe four feet deep and dug by hand into the coral substrate. At some point, it had been used to draw water from the *lens* (layer) of fresh water that lies beneath the island's surface. There were a number of these hand-dug wells around the island. As we continued, it became clear that, although now uninhabited, Swains Island had a lot going on in the past.

I could see flashes of how, under very special circumstances, Alex's ideas about a sustainable-tourism enterprise might be developed on Swains Island. The same remoteness that may make it attractive to adventurous tourists and naturalists, however, also worked against the idea. An enormous investment would be required to bring about such an enterprise on the island. (Never mind one that would be totally green.) My experience with such schemes on hard-to-get-to islands in the Pacific is that the dreams of profitability are usually just that — dreams. Even at very famous Midway Island, only 1,200 miles from Honolulu, with a working airfield and infrastructure, the tourism enterprise proved not to be sustainable. For the time being, I kept these thoughts to myself.

We finally arrived at the old Jennings family residence that was built by Alex's great-grandfather, Eli H. Jennings, in 1929 *(Picture 7.7)*. Both Alex and his brother, David, had lived in the house when they were young children. The old house had mostly collapsed, but the advance party had done a little cleanup, both around the house and the nearby family cemetery. In the plantation period, Swains Island must have been quite the exotic place, with more than 100 Pacific islanders working the plantation. About 800 acres of coconut trees had been planted and cultivated on the island. In addition, its remoteness must have always had an attraction of sorts, bolstered by the unusual freshwater lagoon in the endless salty seascape of the South Pacific. Whatever it was, even the author and poet Robert Louis Stevenson had visited the island during its heyday in the 1880s.

What Is a Tupua? Alex gave us an overview of the remains of the residency and Etena and explained how, as a boy, the island had been his playground. He then asked if we wanted to go to see the *tupua*. Tupua? I had never even heard the word. The advance party had cut a narrow trail to it, and it wasn't far. Togiola explained that a tupua was a sacred site or shrine dedicated to a local spirit in the old Polynesian culture. Alex knew nothing about why it was there, who had built it, or what it meant. He only remembered seeing it when he was a boy.

A short walk through the jungle and there it was, sitting all alone and surrounded by heavy foliage (*Pictures 7.8 and 7.9*). The advance party had partially cleared a small area around it. It looked ancient — like something left behind by a people long gone. It was a small, six-sided (but almost rectangular) platform raised perhaps a foot off the jungle floor, constructed of cut beach-rock slabs, and filled with pebbles and shells. Like welcoming arms, two rows of beach-rock slabs extended out about five or six feet from one side of the platform. Between the arms stood what appeared to be a thick rock pole, perhaps four feet tall. On closer inspection, it was a short vertical column of elongated beach rock or coral. I had never seen such a site. Alex said that the column once had a head of some sort atop it, but it had been knocked off years ago and thrown somewhere in the jungle.

It was really hot, even in the shade of the jungle, but we stood there for some time discussing the tupua. As a Westerner knowing little about Polynesian beliefs and practice, I felt it had to have some practical value or purpose. At the time, we weren't even sure exactly where we were on the island, aside from being on the other side of the lagoon somewhere in the jungle and not far from the old residence. Could the tupua have been seen from the sea in the past and possibly had some navigational purpose or been a marker for something? No one really had a clue. The only thing we knew was that it was close to the very shallow (a few feet deep) far end of the lagoon. (The opposite side of the lagoon from the village, where we had come ashore, was about 50 feet deep.) The tupua remained a mystery as we walked back toward the residence area.

At this point, all of us were overheating. Rather than go back to camp on the other side of the island the same way we had come — through the steamy and insect-filled jungle — we headed directly to the coast, and to the open air and sea breezes. It might have been a little longer way back, but we could walk in the water and cool off. In fact, we had to walk in the water anyway in a number of places, because there wasn't much of a beach. It struck me how different the beach was on this side of the island in comparison to where we had landed. With its steep dune-front and abundant sand down to the water's edge, the beach was obviously on the windward side of the island. I made a mental note and splashed along with Togiola, Alex, and the others.

Back at camp, we doused ourselves with buckets of fresh water from the old water tank, then returned to the tents to relax before dinner. The Jennings family had prepared a large meal for everyone to be served under an open-air, pavilion-like structure in which lights had been hung. Step out of those lights at night and it was pitch black. If we had been back on Tutuila, a Samoan *fale* (a communal thatched-roof structure) would have served this purpose. There probably had been one on the island in the past. I had attended countless feasts and *ava* ceremonies in Samoan fales (*Pictures 7.10 and 7.11*). An ava ceremony is a Samoan custom involving the sharing of an ava beverage to mark important occasions. The same ritual is known as a *kava* ceremony in Hawaii.

Paul Chetirkin and I lit up cigars as we relaxed on coconut tree logs at the tent line. We took in the sea breezes and setting sun, ruminating on what a great day it had been and how incredibly fortunate we were to be on Swains Island. We watched the marine patrolmen and others piling the last coconuts onto the "coconut mountains" they had built on the beach. As the sun went down and darkness fell like a hammer, they lit them just as the call came for dinner. When we returned after dinner, the mountains were ablaze, illuminating the beach around them as if it were daylight. The flaming mountains were so large, we joked that they could easily be seen from the Space Station.

It's Food Poisoning. Dinner was a feast prepared Samoan-style. Samoans are experts at preparing community feasts, which reflect a central tenet of their culture — sharing. The Jennings family brought everything from Pago Pago but also prepared the day's catch from the reef. There was only one downside to the evening: My friend Togiola, the governor, got food poisoning. He was the only one to eat a certain fish caught that day. It is not uncommon for a reef fish to acquire a naturally occurring bacteria or toxin harmful to people. Most likely, Togiola's food poisoning was caused by *ciguatera* — a toxic organism called a dinoflagellate that lives on reef algae and becomes ever more concentrated as it moves up the food chain to fish.

Togiola had eaten one of those fish. He had offered me some of it; thankfully, I declined. He got ill even before we returned to our tents for the night. At first, we thought he must be suffering from dehydration, given the day's exertions, but soon it was clear that something else was going on. A Belgian ER doctor on temporary assignment in American Samoa, who accompanied us to Swains Island, thought the best thing was to simply let it run its course (*Picture 7.12*). I once contracted a case of food poisoning in Melbourne, Australia, and it laid me low for almost a week.

It knocked Togiola pretty low, too. That night, Togiola's chief security officer, Alapati, and the marine patrolmen kept watch over the governor and slept in the sand outside his tent. It was a surreal scene to witness, with the glow of the burning coconut mountains reflecting off these stout Samoans standing guard over their chief. I remember thinking it could be a scene from a thousand years ago on another beach somewhere in Samoa. Fortunately, Togiola felt a little better the next morning, though my dear friend didn't feel like himself again until after we returned to Pago Pago. I was sure glad I hadn't taken a bite of that fish.

I am ordinarily an early riser, but I got up especially early the next morning in the gray light before the stars disappear and the sun cracks the horizon to the east. I was worried about Togiola and went to his tent to check on him. The Samoans guarding Togiola's tent still slept in the sand, and all was still except for the sound of the waves on the reef. Alapati sat up as I peeked into the tent. Our eyes met and we both nodded: The governor was sound asleep, and all was well.

A Revelation. It seemed I was up before most everyone else on the island, but I thought I heard the low murmuring of voices further inland. In search of coffee, I walked to where the Jennings family kitchen was set up. Captain Wally sat in the little kitchen area, where a pot of coffee bubbled on top of a propane burner.

He waved me in to sit down. I think Captain Wally was the senior-most Jennings family member who made the trip. He was a big, overweight man with a full head of white hair and tanned skin. I thought him to be around 70 at the time, but he was actually quite a bit younger. Everyone referred to him as Captain Wally because he was a licensed ship captain and most often the captain of the M/V *Sili*, American Samoa's government ship. Captain Wally had logged many hours at sea and was probably the most well-known ship captain in American Samoa. I had met him a couple of times before in Pago Pago with Alex in discussions about Swains Island, and we had talked a little on the cruise out.

That morning it was just Captain Wally (*Picture 7.13*) and me, sitting in the kitchen area drinking coffee and yarning. (My Polynesian friends also use the phrase "to talk story.") Captain Wally did most of the talking, telling me stories of the family, Swains Island, and the lagoon. He leaned in toward me, resting one arm on a knee, as he spoke. It seemed he either wanted to make sure I was listening, or ensure that nobody overheard him. He told me a lot of what I already knew, especially about the lagoon. One story was how they used to place coconut logs in the lagoon for about a year before using them; and how, when they were taken out, they were so hard you couldn't drive a nail through them. He said repeatedly, "The lagoon — it petrifies things!" What he meant was that the brackish water mineralized the logs. Then he dropped a bomb that changed everything, and put me on the path that would lead to the Second Expedition.

He said: "My grandmother she told me her mother told her that there had been a channel into the lagoon. Yeah, a channel! But big hurricane — maybe hundreds of years ago — come and close it." While Captain Wally talked, he pointed to the shallow end of the lagoon across the island, where we had been the day before and seen the tupua. It was as if Earth stood still for a moment, and everything I had seen and heard on the island came rushing together in my mind. What Captain Wally's grandmother had meant by a "channel" could have been a lot of things. It simply may have meant that canoes were able to float over the coral rubble into the lagoon at high tide. Whatever she meant, however, the bottom line was that canoes might have somehow entered the lagoon in the distant past. Swains Island would have been a perfect place for Polynesian voyagers to stop and replenish their food and water, and to find materials to repair their voyaging canoes. It also would have been the perfect place to sit out rough seas or storms. The lagoon would have been what we call a perfect "hurricane hole": a place to hide during a storm.

It was then that my imagination began to work overtime. In the past thousand years of Polynesian voyaging, could it be possible that a sailing canoe might have entered the lagoon and never made it out? Or that unsuccessful repair efforts had left the remains of a sailing canoe in the lagoon? It is known that the great Polynesian voyaging canoes, which carried early Polynesian seafarers from Samoa to and from distant islands across the Pacific, disappeared from the Pacific around the 1400s. Little or no physical remains of these historic vessels are known to exist. Was it possible that an unexpected opportunity of discovery might lie at the bottom of the lagoon at Swains Island?

What mysteries might the lagoon hold if some form of mineralization had been at work in the past (as it is today) that could preserve coconut logs and other wood? While we didn't know the exact location of the tupua, it was near the very shallow end of the lagoon. Could its purpose have been to mark the "channel," or the way into the lagoon? The tupua was certainly from an earlier time and was a genuine Polynesian artifact. You know the old saying: "A little knowledge can be a dangerous thing." That is exactly the position I found myself in as we boarded the *Lady Naomi* for the sail back to Pago Pago.

On our return voyage, I tested my hypothesis with a number of Swains Islanders onboard to try and find out what they might know about the lagoon or may have heard through family stories. No one on the ship really added much to the notion that the lagoon might hold secrets. The consensus was more like, "Yeah, why not? Or could be?" I don't think anyone had given much thought to past Polynesian voyages or what might be in the lagoon. The idea stuck with me, however, and I would have to enlist others far more knowledgeable than myself if anything were to come of it. The initial seeds for a Second

Expedition to Swains Island had been planted, though, and two years later, they blossomed into Jean Michel Cousteau's award-winning documentary film, *Swains Island — One of the Last Jewels of the Planet.*

Chapter 7 Pictures

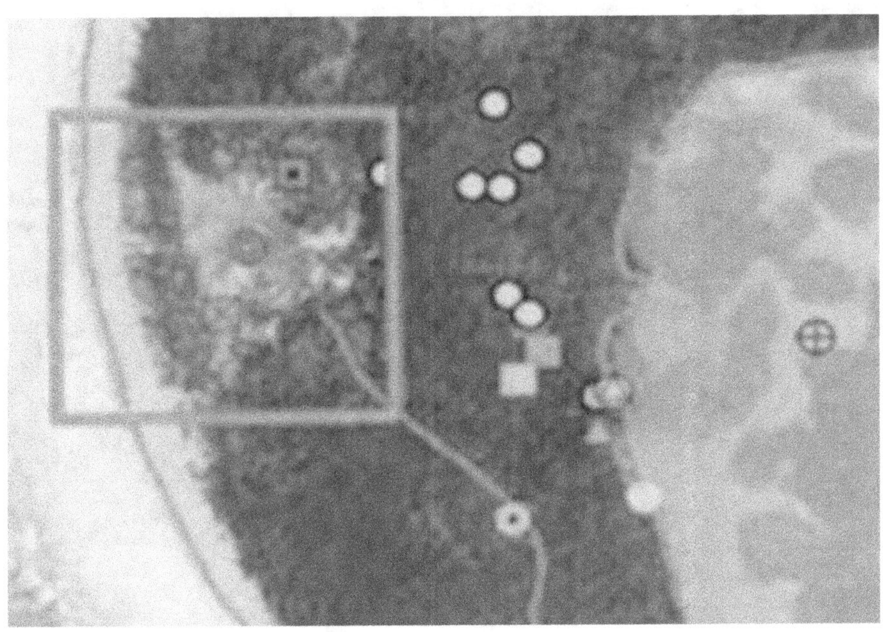

Picture 7.1. The Taulaga village area is within the square. The line is the road to the old Jennings family residence.

Picture 7.2. The old church.

Picture 7.3. The water tank, still in use.

Picture 7.4. The lagoon as seen from the end of the foot path. This photo is from the *Hokule'a* visit a few years after our expedition. Nainoa Thompson is, I believe, the man standing in the lagoon.

Picture 7.5. The old road.

Picture 7.6. The Jennings family cemetery.

Picture 7.7. The front of the old Jennings residence at Etena.

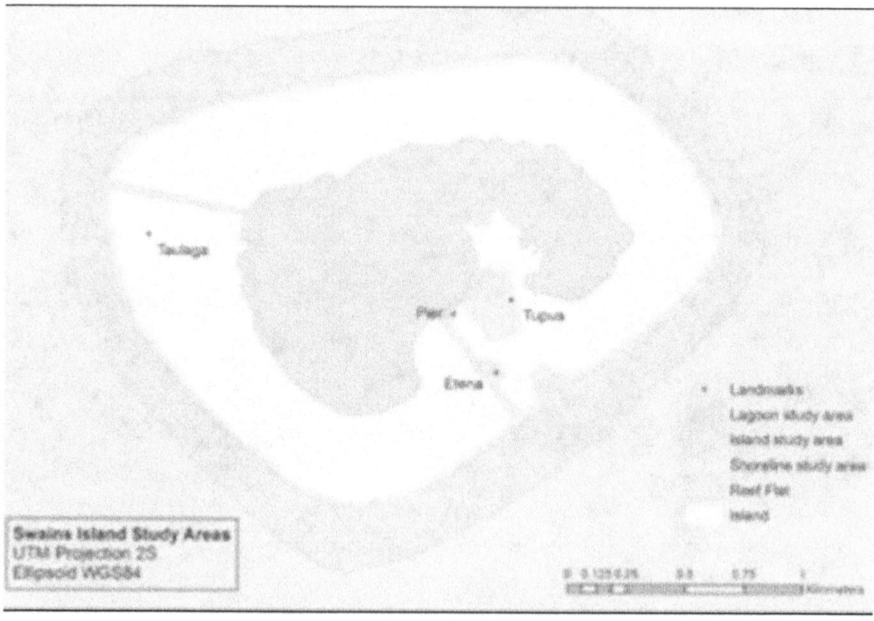

Picture 7.8. Location of the tupua.

Picture 7.9. The tupua.

Picture 7.10. Preparing the ava beverage for an ava ceremony.

Picture 7.11. The author accepting the ava cup during the ceremony.

Picture 7.12. The good Belgian doctor, *left,* and Governor Togiola Tulafono on the *Lady Naomi* during the voyage out.

Picture 7.13. Captain Wally enjoying the ride on the *Lady Naomi*.

Chapter 8

Interlude

As the months progressed, it seemed like we were characters in the European folk story about making stone soup. I wondered many times if the Second Expedition would turn out at all. Sticking with the old proverb, "Where there's a will, there's a way," and finding the right individuals to help inspire and motivate others, as well as ourselves, had made the difference.

Many things affected the undertaking of a Second Expedition between the time when the First Expedition returned to Pago Pago and the Second Expedition departed from the same dock two years later. What follows is a summary of the events and interactions that I was part of, and I think they are important or interesting aspects of the story. Other things were also happening at this time, and I am sure some of my colleagues would have interesting stories of their own to tell as the Second Expedition began to take form.

After arriving back in Pago Pago, I had little time to further explore the idea of another expedition to Swains Island. We found out what we went there to learn, and then some. So what value would be added to justify another expedition? As was always the case when I was in American Samoa, I had to get as much out of each visit as possible. Consequently, there were always too many things planned and scheduled. The trip to Swains Island had been no different. For example, another important purpose of the trip was to initiate a swimming program for American Samoan youth (in fact, for all Samoans). I had been shocked to learn that a major cause of death among Samoan youth was drowning. Most Samoans did not know how to swim. Nancy Daschbach first brought this to my attention, but the question was: What could we do about it? When we returned to Pago Pago from Swains Island, a world-class long-distance swimmer, Bruckner Chase, awaited us to help our cause.

Bruckner is a celebrity in the long-distance swimming world, and also an extraordinary and compelling spokesperson. We had recruited him to make highly publicized long-distance swims in National Marine Sanctuaries to bring attention to our events and programs. His longest swim for us had been 27 miles across California's Monterey Bay from Santa Cruz to Monterey. In American Samoa, he was to swim seven miles, from the nearby island of Annu'u to the beach at Utulei, near Pago Pago, to help focus attention on swim training (*Picture 8.1*). He did far better than make a promotional swim, though — he initiated a swim program that eventually reached 5,000 American Samoans. After Bruckner's swim and the events surrounding it, I was back on planes to make my way home to Washington, D.C.

Testing the Idea of a Second Expedition. One of the first things I did upon returning home was to contact Jean Michel Cousteau. I briefed him on what we had learned at Swains Island, and told him of my half-thought-out idea about the mystery of the lagoon. At this point, I think he was more interested in the diving we had done and the marine ecosystem surrounding the remote Pacific island. I agreed to keep him informed about further developments in American Samoa.

Not long after speaking with Jean Michel, I contacted my good friend Nainoa

Thompson in Hawaii and ran by him the proposition that Polynesian voyaging canoes could have stopped at Swains Island in the distant past. Nainoa is a native Hawaiian, known throughout the Pacific as "The Navigator." It was Nainoa, probably more than anyone else in Hawaii or anywhere in the Pacific, for that matter, who helped resurrect the great Polynesian sailing tradition that colonized the Pacific. His voyages on the *Hokule'a*, a double-hulled traditional sailing canoe, had driven a cultural renaissance in Hawaii and elsewhere (*Picture 8.2*). I had met him after his reputation as The Navigator and captain of the *Hokule'a* was already well established (*Picture 8.3*).

Nainoa Thompson. Nainoa has a style unto himself. He is quiet by nature and often seems laid back, but he is always listening. He has a way of seeming to be the student, even when he is the teacher. He is a passionate speaker, and when he speaks, he burns with a contagious intensity. I think it was probably this very natural way of his that endeared him to one of the last "ancient navigators" of the Pacific — Mau Pialug. From the island of Satawal in Micronesia, Mau had for decades taught others the ancient ways of sailing. Nainoa became his pupil in the 1970s; the rest is history with the *Hokule'a* voyages across the Pacific and, later, around the world. A good book written by Sam Low, *Hawaiki Rising: Hokule'a, Nainoa Thompson, and the Hawaiian Renaissance*, published in 2013, tells Nainoa's story. It also tells something of Mau's history — and is well worth reading.

I first met Nainoa one evening at his parents' home on Oahu. His mother, Laura Thompson, or "Auntie Laura," as I came to know her, was a member of the citizens advisory council supporting the idea that the Northwest Hawaiian Islands should become a National Marine Sanctuary. (In 2006, President George W. Bush declared these islands the Papahanaumokuakea National Marine Monument, the largest marine protected area in the world at that time.) After many council meetings, Auntie Laura and I found ourselves on the same wavelength about a lot of things. We had established trust between us. One evening, she invited me to her home in the Niu Valley to meet her husband, "Pinky," and her son, Nainoa. This began a long relationship with

Auntie and her extended family. I remember that first evening well, sitting in Auntie's kitchen and just talking. Pinky was a teacher and a revered leader in the Hawaiian renaissance movement. He was also, as I came to realize, Nainoa's North Star. Pinky casually leaned in the doorway as we talked. He seemed to be a man of purpose and comfortable in his own skin and beliefs. I cannot remember exactly what we talked about, only that it was not idle chatter. (I don't know if Pinky ever engaged in idle chatter.) Nainoa mostly listened to the conversation that night. I was probably talking too much anyway. From that evening forward, I almost never made a trip to Hawaii that I didn't go up the Niu Valley to visit the Thompson house.

What little I knew about Polynesian voyaging before the First Expedition to Swains Island, I learned from Nainoa. We spent many hours talking about voyaging, about the *Hokule'a* and its history, about navigating the stars, and everything else imaginable. I even sailed a little with him on a voyaging canoe. We gradually became real friends. In the end, we found we were both passionate about many of the same things. I like to think our friendship is what Auntie Laura had intended from the very beginning.

I learned from Nainoa that few artifacts or written records remain about ancient Polynesian voyaging canoes. In fact, the building of the *Hokule'a* itself had been the result of almost a decade of research and experimentation by archeologists, anthropologists, and interested individuals. They had called it "experimental archeology." Therefore, if the remains of a voyaging canoe might lie "petrified" at the bottom of the lagoon on Swains Island, discovering it would be an enormous find for all of Polynesia. Nainoa reasoned that early voyagers must have visited Swains Island on any number of occasions. No one really knows all the routes, nor the frequency, of Polynesian voyaging across the Pacific, but Nainoa suspected that far more voyages had taken place than it is possible to know. This was all I needed to feed my imagination and to continue exploring the possibilities for a Second Expedition to Swains Island.

Hans Expands the Scope. My next stop to discuss the idea further was with

Hans Van Tilburg. Hans (*Picture 8.4*) had written the original background report on Swains Island a few years earlier. Not only was Hans a consummate marine archeologist, but he was also experienced in field expeditions. Consequently, Hans and I had many discussions about what a Second Expedition might look like and how it could be conducted. Unlike other ocean exploration/discovery projects I had been on, exploring the possibilities of the lagoon seemed a pretty straightforward undertaking. We had access to side-scan sonar and the experienced personnel needed to map the bottom of the entire lagoon. The deepest part of the lagoon was only around 50 feet; it would be easy to scuba dive on a target and determine what it might be. Once we got to Swains Island, it would all be so easy. (As usual, this did not turn out to be the case!) The expedition would have to be an "official" NOAA operation, and that would affect how dive operations would be conducted.

Hans, always the archeologist, proposed that if there were going to be a Second Expedition to Swains Island, it should expand beyond the mysteries of the lagoon to include a maritime cultural heritage resources survey of the island. He reasoned, given all it would take to organize another expedition, that this might be the only opportunity to appropriately survey the island. This made perfect sense on several levels, especially given the new direction of NOAA's Office of National Marine Sanctuaries Maritime Heritage Program, which was already undertaking similar surveys elsewhere.

Hans was right! I give him full credit for moving us toward undertaking the most thorough cultural heritage resources survey possible. In my narrower perspective, I was overly focused on the mysteries of the lagoon. That was the story I thought would attract the attention needed to embrace a Second Expedition and engage a large audience. I still think this is true, as it gives tiny Swains Island a mystique and distinguishes it from other islands. Whatever mysteries the lagoon might hold, however, they are tied directly to the history of the island. Hans had earned himself a new assignment: design a plan to conduct a scientific cultural resources survey at Swains Island, especially in and around the lagoon. It would take a number of iterations to put together a feasible plan, given the difficulties of getting to and surveying the place. Compared with what the Second Expedition eventually became, the first one seemed relatively easy.

Would a Second Expedition Be Worthwhile? Exploring the mysteries of the lagoon was tantalizing in its own right, but it was not enough to warrant a Second Expedition, and neither was a cultural resources survey of tiny, faraway Swains Island. The benefit to American Samoa had to be greater, especially given the limited resources. *A Second Expedition to Swains Island would have to be a means to an end to bring attention to, and promote, American Samoa — or it wasn't worth doing.* The truth is American Samoa is so far removed from the U.S. mainland, most Americans don't even know it is a U.S. territory. Nonetheless, if a documentary film could be made to reveal the hidden mysteries of a remote and "deserted" island, it might attract enough attention to provide significant value to American Samoa. *The timing proved fortuitous because momentum was building to expand the tiny Fagatele Bay Sanctuary. A Second Expedition to Swains Island could inform our view of how large of an expansion might actually make sense.* The logic seemed to fit. Now all I had to do was pitch the idea based on our discoveries from the First Expedition and a compelling mystery to solve. This proved not to be easy.

I pitched the documentary film idea to a number of places as well as to a few filmmakers I knew. They thought the idea was fascinating, but so were the size of the budgets and the projects they envisioned. Perhaps it was my inability to explain what I was attempting to pull together and achieve. The cost and control of their production models just didn't fit the situation. It soon became clear that only one person could make such a film, and he already knew about Swains Island. It was my old friend, Jean Michel Cousteau, of course. I think Jean Michel agreed to make the film because of our friendship and the missed opportunity the year before. It certainly wasn't for the money because there wasn't any. The resources that could be scraped together would be needed to carry out the basics of the expedition. Jean Michel took no salary or fee. All I had to do was get him and his associate, Jim Knowlton, who also took no salary, to Swains Island and cover their expenses. They would design and shoot the film and do all the post-processing and editing for pennies. Jean Michel and his foundation, the Ocean Futures Society, would own the film — and that was fine with me.

I had known Jim Knowlton for a few years and piqued his interest in Swains Island during NOAA's Ocean for Life Program, held earlier that summer. The Ocean for Life Program was one of my creations after 9/11. It brought together kids from the U.S. and the Middle East to promote cross-cultural understanding through their greater appreciation of the ocean and how it connects us all. Jim was the director of the Jean Michel Cousteau Media Camp for Ocean for Life that summer (2013), which turned out to be the last year of the program. Jean Michel and Jim were supporting Ocean for Life when they signed on to film the Second Expedition to Swains Island. I don't think either of them could resist the temptation to join the expedition. The pieces were starting to come together; if the Jennings family and the territory of American Samoa agreed to our return, that is.

The Pieces Come Together. The crucial piece to completing the puzzle was to enlist the Jennings family. An expedition would be impossible without them. On a trip to Samoa, I proposed to the family the idea of a second science expedition to, and documentary film about, Swains Island. There was genuine excitement, especially when people heard that Jean Michel Cousteau would be making the film! They had complete trust that whatever he created would do well by Swains Island. The Jennings family was also pleased that Hans Van Tilburg would be involved. They were impressed by the study he had done a few years before, and now knew and trusted him.

Unfortunately, during this period, Alex Jennings was off-island most of the time. It turned out he would not be able to make the upcoming expedition. Alex is a commercial pilot and was under contract, busy flying to make a living. This is when his brother, David Jennings, stepped up and became an essential partner in the planning and execution of the Second Expedition.

David has a way about him that makes people want to be in his company and listen to what he has to say. Partly it is his broad smile, but there is also something in his eyes — they glisten. David is, by nature, a critical thinker and problem solver. Early on, I thought he was a little suspicious of me and guarded when we first met in Pago Pago prior to the First Expedition. That expedition, in which he had not participated, convinced him that we really were all about doing things to benefit the people of Swains Island and American Samoa. So as planning moved forward, David was all in, and even led an advance party to the island. His Navy career had taught him a lot, and he brought all of those skills to the expedition. It turned out that he also plays the guitar and sings, which would make for many pleasant and memorable evenings.

Hans's plan for the Second Expedition required a science party of eight to ten individuals who would be supported on the island for as many as eight days. Then, in addition to those who would support the science party, still others were penciled in, including some who could be on the island for only three days due to schedules. The latter group revolved primarily around Jean Michel and me. We had to return to Pago Pago after three days to fulfill other responsibilities, such as a reception with prominent Samoans to honor visiting dignitaries (primarily Jean Michel; I was an old hand by then), and then to help celebrate the one-year anniversary of the new Ocean Center (*Picture 8.5*). Consequently, at least three roundtrip voyages would have to be made to Swains Island: one to put David Jennings and the advance party ashore and return; a second to bring the main party to the island and return to Pago Pago after three days; and a third to bring back the science party and the support staff.

In early 2013, Hans completed the final, 26-page plan, which I approved. It was clear that the logistical requirements would be far greater than they had been for the First Expedition. This included new requirements to support a host of computers, sensing equipment, and more. For example, a portable re-compression chamber would be set up on the island to support dive operations. The territorial government of American Samoa and the Fagatele Bay National Marine Sanctuary also had important roles to play and critical resources to contribute to the expedition. My friend, Governor Togiola Tulafono, could not make the Second Expedition, but he made sure the territorial government did whatever was necessary. Gene Brighouse and her staff *(Picture 8.6)* at the marine sanctuary again became the glue that held all the pieces together.

As the months progressed, it seemed as if we were characters in the European folk story about making stone soup. A wandering soldier enters a starving village carrying a magic stone. He places the stone in a large cauldron and declares it will make soup for the village, only someone has to find some carrots, someone else a potato, etc. A wonderful soup is made and the village is fed. I wondered many times if our plans would turn out at all. Sticking with another old proverb — "Where there is a will, there is a way" — and finding the right individuals to take part in inspiring and motivating others, as well as ourselves, made the difference.

I designated Hans as chief scientist of the expedition's maritime cultural survey. There could be no other choice, and I had complete confidence in his knowledge and capabilities. Still, it was a tall order for anyone. He first had to attract experienced researchers/adventurers to the expedition. Members of the party had to be self-reliant, comfortable without a lot of frills, and experienced camp mates. They would be camping on a remote, deserted island, alone in the South Pacific, with the closest support more than 200 miles away in Pago Pago. For a good part of their stay, no ship would be standing by to help them or transport them to safety. The chemistry among them had to be right.

I made a point of approving everyone proposed to make the trip, except for the Jennings family members and their guests. Of course, everyone who Hans selected measured up. We both are of the same mind on such matters. I wanted to know as many of them as I could. Those I hadn't met, or didn't already personally know, I made a point of meeting on the voyage out to Swains. I would have had no problem scratching anyone I didn't think was solid. Hans's leadership style is "light hand on the tiller," which worked perfectly. The party of volunteer researchers he recruited came together seamlessly under his leadership.

Finding a Ship. There was, however, still one missing piece: coordinating the complexity of ship operations, small boat and dive-support requirements, and myriad overlapping technical and logistical details. I needed someone in whom I had complete trust to oversee the operational details of an expedition to such a remote place. The task required an individual with expertise and experience across all details of such endeavors. For this role, I turned to Dana Wilkes, the chief mariner of NOAA's Office of National Marine Sanctuaries. In my mind, a more qualified and trusted person did not exist. Dana is a graduate of the U.S. Naval Academy, a former naval officer, and a NOAA Corps Officer. At the time of the Second Expedition, we had been working together for almost 10 years.

I first met Dana Wilkes (*Picture 8.7*) more than a decade before on the NOAA ship R/V *MacArthur*, when we had been struggling with launching and operating small, manned submersibles from NOAA vessels. Nancy Foster, then head of NOAA's National Ocean Service, had put me in charge of these NOAA operations. Over a very dangerous two-year period, we successfully conducted complex operations on five vessels. This had never been done before on a NOAA ship, nor has it been done since. In this crucible, I quickly recognized Dana's extraordinary qualities and abilities. He was simply indefatigable, and I considered him key in making that entire program work — and work safely. Consequently, when I was put in charge of the Office of National Marine Sanctuaries, I pulled him in as soon as I could.

By time of the Second Expedition, Dana had already changed the seagoing culture of National Marine Sanctuaries. He was responsible for the design, construction, and operations of a new small-boat fleet to support National Marine Sanctuaries around the country and in the Pacific. As a NOAA Corps officer, he had also served in NOAA's diving office and was an experienced dive officer. Most importantly, Dana was also very familiar with Samoa and Samoans. He knew them well, and they knew and respected him. There could be no mistakes at Swains Island.

The M/V *Sili*, the only seagoing vessel operated by the territory, was also the only ship available for the expedition *(Pictures 8.8 and 8.9)*. It would take us back-and-forth to the island. When I mentioned this to Dana, he chuckled and said, "Are you kidding? I thought it was still rusting away, waiting for major repairs." He obviously knew the M/V *Sili* and was less than impressed.

As it turned out, the M/V *Sili* still holds the honor of being among the most worn-out and decrepit ships I have sailed on. Dana had been right to chuckle when I mentioned the vessel to him. Acquired by the territory in 2000 and already old, she had been an oil-rig supply ship, common in the Gulf of Mexico. Having been specifically designed for that purpose, she was almost all open deck behind her small, multi-level deck house, which was situated forward near the bow. She had been a heavy-duty work ship with no built-in passenger accommodations. As a result, a few modifications were made to the M/V *Sili* when she arrived in American Samoa. Benches were installed under awnings on both the main and upper decks behind the deck house to carry some passengers in the open. The M/V *Sili* was the only ship the territory had within its control to move cargo around the American Samoa archipelago, nearby Samoa (formerly Western Samoa) and other close by islands. For all her time in American Samoa, she seemed to have been run on a shoestring. Certainly, a lot of maintenance had gone by the board over the years. It was even difficult, at times, to find a competent crew to man her.

Fortunately, Dana was very familiar with the M/V *Sili* as well as the Pago Pago Port Authority and the U.S. Coast Guard officer stationed at the port. Everyone kept an eye on the ship while the territory continued to search for a replacement vessel. Dana was, in fact, helping the territory clarify their needs and requirements for a new vessel. Nevertheless, the M/V *Sili*, with a pick-up crew primarily consisting of Fijians and with Captain Wally at the helm, managed to make the round trips to Swains Island. She was crucial to the expedition: no M/V *Sili*, no expedition. Those of us who sailed on the *Lady Naomi* on the First

Expedition had been spoiled. Our voyages on M/V *Sili* bore no resemblance to those earlier trips. At times, we had wanted to close our eyes and have the *Lady Naomi* magically reappear, as if we had awoken from a nightmare.

The Day We Finally Sailed. I arrived in American Samoa with Jean Michel Cousteau, Dana Wilkes, Jim Knowlton, and a few others the night before the M/V *Sili* was to sail. It had been a long flight, and even though it was near midnight when we landed at Pago Pago, the governor was there to meet us at the airport and expedite our entry process. It was good to see my friend Togiola and other familiar Samoans. True to their traditions, they made Jean Michel and everyone feel at home as honored guests. The M/V *Sili* was scheduled to depart at 11 a.m. the next day. Hans, David, Gene, and others had been at work for days, organizing everything and everyone to go aboard. A large shipping container was secured to the open rear deck of the ship, into which everything was carefully loaded for the journey. All we had to do was get aboard the ship in time to sail. The hard work over the past two years had paid off. At the airport, I could see the excitement in everybody's faces, in their eyes especially, and in their body language. It was a great feeling that, after two years of struggle, the expedition would leave the dock the next day, and this time, Jean Michel would actually be on board.

That morning, as per usual when in American Samoa, I got up at 5 a.m., which was easy given the time change. I made my way to the only gym in Pago Pago for an early morning workout. I would do this every day. Gene also worked out early. Most mornings, she picked me up in the dark and took me to the gym. She is an excellent racquetball player and played with a group. Lelei was also in the racquetball group and was there most mornings. I think Gene was the best player among them. I also played with her once or twice and was beaten handily. The morning of the M/V *Sili*'s departure, Gene was ironing out last-minute details, so Dana Wilkes came to pick me up. Dana and I had previously worked out together in American Samoa, and had our own routines.

Dana arrived at 5 a.m. on the dot in the pitch dark. I stayed at a spectacular place — the Moana O Sina (*Picture 8.10*) — outside town along the coast. The lodge had become my home away from home over the years when I was in Samoa. The owners had become friends and would always set up the coffee machine the night before so I could make a quick cup in the early morning before heading to the gym. The main building and residence at Moana O Sina rise up maybe 15 or 20 feet above the dirt-road parking area. A fairly steep set of lava rock stairs goes up to the building and terrace.

I was finishing my coffee that morning when I heard Dana come in. We talked in whispers because everyone else in the house was, of course, asleep. We headed out into the darkness, onto the terrace, and down the lava rock steps. One second we were side by side, and the next, I watched Dana tumble down the steps. Somehow he remained on his feet, but even so, he twisted his ankle badly. Weeks later, we learned that he had actually broken it. We still went to the gym that day, and Dana limped around, doing his routine. When we returned to Moana O Sina, I could see he was in great pain. Fortunately, I had some heavy-duty painkillers in my medical kit, and he reluctantly took one. "See you at ship," I said as he drove away. That Wilkes is a tough hombre, but I didn't know quite how tough until I got to the ship. Still limping, Dana was on and off the ship, checking this and that, and chatting to this one and that one. His ankle was quite swollen, but he stuffed his foot into his boot and laced it tight. Those boots were not coming off for a couple of days, and he was going to make this cruise no matter what. I gave him the rest of the pain pills.

Gene, who was making the trip to Swains Island for the first time, had more than earned it. Another tough individual, she took her motion-sickness medication before stepping aboard. She was so sensitive she almost got seasick just looking at water, but she had to make the trip, regardless. A few others, who had been invited by the Jennings family, also filed aboard. It was a once-in-a-lifetime opportunity for them to participate in such an adventure. As we boarded the M/V *Sili*, I could already see the look in Jean Michel's eyes. They spoke volumes to me, and it all seemed to be in French, as he looked over the ship. Clearly, he was none too pleased. To pre-empt what he was about to say, I quickly uttered, "Well, it isn't the *Calypso* — that's for sure." He gave me a look! As the M/V *Sili* came off the pier and got underway, her condition, or lack thereof, was all but forgotten. After two years of struggling and planning, the Second Expedition to Swains Island had begun.

Chapter 8 Pictures

Picture 8.1. Bruckner Chase, *right*, arriving at the beach at Utulei after his seven-mile swim.

Picture 8.2. The *Hokule'a* voyaging canoe off of Swains Island during the 2014 Worldwide Voyage.

Picture 8.3. The author, *left*, and Nainoa Thompson reunite in early 2016, when the *Hokule'a* visited Washington, D.C., during its Worldwide Voyage.

Picture 8.4. Hans Van Tilburg at the time of the Second Expedition.

Picture 8.5. The Tauese P.F. Sunia Ocean Center in 2012, prior to its opening.

Picture 8.6. Gene Brighouse, *third from right, bottom,* and key staff at the Ocean Center in 2012. The author, *fourth from right, bottom,* stands next to Nika, a high chief from Western Samoa.

Picture 8.7. The intrepid Dana Wilkes.

Picture 8.8. The M/V *Sili* underway. Paint can be deceiving!

Picture 8.9. The M/V *Sili* tied up at Pago Pago.

Picture 8.10. The view looking seaward from Moana O Sina lodge, the author's favorite place to stay on Tutuila.

Chapter 9

A Stormy Night

As I pushed open the hatch and stepped onto the deck, the glare of powerful flood lights high in the deck house illuminated the back deck. Water rushed across the deck as waves broke over the port side of the ship. At least six to twelve inches of water streamed under the benches. I thought it wouldn't take much for a big wave to hit the benches — and the people on them — behind the deck house.

As the M/V *Sili* glided slowly down the channel to the open sea, most of us on board stood along the rails or at the bow of the ship (*Picture 9.1*). The green hills that rim the harbor around Pago Pago never looked more beautiful as the sun glistened upon them. It was a later departure than planned. Well-wishers on the ships and boats we passed and standing along the shore waved at us, and we at them. Everyone in Pago Pago knew the M/V *Sili* and where we were headed. It was a great a feeling to leave on an expedition with such an enthusiastic send-off.

Into the Open Ocean. The ship exited the harbor, moved into the open sea, and passed the small island of Annu'u, from which Brucker Chase had made his swim two years before. The bow of the M/V *Sili* then turned almost directly due north toward Swains Island. American Samoa and Pago Pago disappeared astern as the ship's wake straightened out. All of the passengers and members of the science party began to find places on the benches behind the deck house, where they planned to spend the night. At this point, the cruise seemed almost like any other, except for people sleeping on the benches and on the deck. Water sloshed along the side of the ship, as it gently swayed, moving through the sea. On this old ship, it was easy to feel the engine vibrate throughout the vessel, and there was the ever-present smell of diesel fumes.

Within the small deck house toward the bow was a commensurately small galley with tables and benches. If I remember correctly, some crummy packaged foods, sodas, and coffee had been provided for the cruise and were available behind a small counter. People who had been on the M/V *Sili* before and knew the deal staked out the few benches and tables in the galley. The small room was easily crowded. Most of the passengers grabbed something from the counter and returned outside to the benches or other places on deck.

The entire scene was thoroughly disorganized and a bit chaotic. Jean Michel (*Picture 9.2*) couldn't help but ask, "Who is in charge?" I must admit, I should have paid more attention when Dana chuckled when we first discussed the ship — not that it would have mattered. Dana, who was standing with Jean Michel, just smiled with a grin that said "I told you so." I said to Jean Michel, "Why, the captain is in charge, of course. But don't forget — this is Samoa, and you have to adjust. Things are done differently out here." Jean Michel knew that, of course, and we all just grinned and bore it, silently acknowledging that it was going to be a long night.

Jean Michel and I were the two highest-ranking individuals on the ship and, according to Samoan custom, we received the best accommodations. We were thus offered the luxury of two bunks in what appeared to be the life-jacket storage locker on the deck above. While piles of life jackets and other things were heaped all around, I think we were the only ones onboard with *actual* bunks. (Captain Wally, however, probably had a bunk behind the bridge on the deck above us.) Jean Michel and I stowed our stuff in the bunks and headed down to the galley. I think my old friend might have been thinking, "What has Dan gotten me into this time?"

A short time later, I walked around the ship to see where everyone had bedded down. Sleeping bags and blankets were spread on the benches and in corners under the awnings. Gene was already sound asleep on a bench on the upper deck, numbed by the motion-sickness medication she had taken (*Picture 9.3*). Members of her staff made sure she was tied in. Everyone seemed to be in high spirits. I caught up with Hans and members of the science party. They didn't fret at all about sleeping on a bench or on the deck (*Picture 9.4*). This would be Hans's first time on the M/V *Sili* and on Swains Island. As we talked, he said, in his usual low-key manner, "It is what it is, and tomorrow we will be at Swains, anyway." He smiled as he said it, with those expressive raised eyebrows of his. That's exactly what I needed to hear, and it's exactly why he was the chief scientist. What's a little adversity and discomfort when we are on an expedition?

Back in the galley, everyone swapped tales, but many simply looked at Jean Michel and hung on his every word. They had never met him, but the name "Cousteau" was like magic. It's possible that the Samoans thought he was actually his father, Jacques. They may even still think so, and tell tales of sailing to Swains Island with Jacques Cousteau. I asked Dana where he was bunking for the night and he pointed to the floor. "Right there on the deck," he said. His ankle was still awfully swollen, painful, and stuffed tightly into his boot, but he said the painkillers I gave him helped. I remember asking him what he would do the next day when we had to wade ashore through the surf. He said he would just keep his boots on! Later, I saw him limping around in sandals. He took his boots off, after all. When I asked how the ankle was, I got a typical Dana answer: "I'm good."

The Storm Begins. About 10 p.m. that evening, as everyone was settling in for the night, the ship started to roll harder and buck a little. We were moving into a storm. Jean Michel and I climbed the tight stairs to our "luxury" bunks and moved some things around to better stow our gear. On this ship and on this night, we decided it was best to sleep in our clothes. I think we each had a pillow and a sheet of sorts for our bunks. After talking a little about Swains Island and the diving we planned to do, we both drifted off.

Then, around 1 a.m., we awoke as we rolled from side to side in our bunks. At times, we had to hold on or chance rolling out of the bunks. We were clearly in the middle of a pretty good storm. Captain Wally expertly navigated the M/V *Sili* along the waves, taking mostly rolls. He had made this passage many times, and he knew the ship and her ways well.

I decided it prudent to check on our people, especially those who were sleeping on the benches and the back deck. Jean Michel looked at me in the dim light and asked, "Where are you going?" "I'm going onto the back deck to check on things," I replied. Jean Michel then smiled and said, "Goodbye. Don't wake me up when you get back." (Like he would fall asleep in the 15 minutes I would be gone!) I made my way down the swaying stairs to the galley. It was dark except for a small night-light behind the counter. Sprawled everywhere were people lying on tables, benches, and the deck. Dana lay in a contorted position, face down on the deck, exactly where he had pointed earlier. I don't know how many people were actually asleep, but I could see a grimace here and there and several sets of eyes pop open and shut. Dana, however, was dead to the world as I stepped over him to get to the hatch to the back deck. Maybe the pain pills had knocked him out.

It was a good thing the hatch was high in the wall on the bulkhead and off the deck. As I pushed it open and stepped onto the deck, the glare of powerful floodlights high in the deck house illuminated the back deck, where I could see water rushing across the deck and waves breaking over the port side. At least six to twelve inches of water streamed under the benches. It wouldn't take much for a big wave to crash over the benches — and the people on them —

behind the deck house. Several people had retreated to the upper deck, but not everyone. Stephanie Grandulla (*Picture 9.5*) sat upright in her sleeping bag on a bench on the starboard side directly behind the deck house, and beamed in the glow of the light as water rushed beneath her. I sloshed over to see if she was okay. She seemed to be having the time of her life! She gushed with excitement as she explained that a large fish had just been swept beneath her bench in the rushing water. She asked me what it might have been. I had no idea about the fish, but I now knew that Stephanie was a good choice to be part of our science party and expedition. No matter what challenges she might encounter, she could handle them.

I hadn't been so sure about Stephanie when Hans and her supervisor, Jeff Gray, superintendent of Michigan's Thunder Bay National Marine Sanctuary, had proposed that she be part of the science party to assist in the expedition's side-scan survey. The survey was to be conducted by Matt Lawrence from the Stellwagen Bank National Marine Sanctuary in Massachusetts (*Picture 9.6*). I knew Matt fairly well and had even dived with him. There wasn't a more competent and solid marine archeologist to do the survey on remote Swains Island. I had met Stephanie only once before she was proposed for the science party. Subsequently, when on a personal expedition to Thunder Bay to dive shipwrecks, I took the opportunity to interview and size up Stephanie for the expedition. Truthfully, I wasn't a hundred percent convinced, but I went along with Hans and Jeff and put her on the expedition. That night on the back deck of the M/V *Sili*, there in the glowing lights, any reservations I may have had about Stephanie washed away with the water streaming across the deck. She turned out to be a great addition to the expedition.

I walked around the decks and checked on everyone. I hadn't seen anyone looking panicky or even overly concerned. In fact, many just looked like lumps on a bench. Gene was still tied into her bench and asleep from her medication. It was exciting as the waves rolled over the side of the ship and around our large storage container, which was firmly secured to the deck. Excitement, however, could easily change to tragedy if anyone tried to cross the deck that night. At that moment, however, all was well, and I made my way back up to the bunk room and Jean Michel. "How is it?" he asked. I think I responded with something like, "I'm glad we have our luxury bunks." Tomorrow we would reach Swains Island, and the storm would be a good tale to tell.

Chapter 9 Pictures

Picture 9.1. On the bow of the M/V *Sili*. *From left to right*: Hans, Jean Michel, the author, and Jim.

Picture 9.2. Jean Michel on the M/V *Sili* as we left Pago Pago.

Picture 9.3. The upper deck behind the pilot house. Note the benches and the tarps to keep weather out.

Picture 9.4. Discussions begin on the way to Swains Island. *From left to right*: a new Samoan friend, Jim Knowlton, Hans Van Tilburg, and the author.

Picture 9.5. Always smiling, Stephanie Grandulla brought great enthusiasm to our Swains Island adventure.

Picture 9.6. Matt Lawrence with the Swains Island lagoon in the background.

Chapter 10

The Second Expedition Lands

Going ashore was essentially the same process it had been when we landed from the Lady Naomi two years before. A single, battered, aluminum skiff came out through the cut in the reef, and everyone and everything had to somehow get ashore on the small skiff. The only difference was that the M/V Sili was not a sleek ferry like the Lady Naomi, with her stern-end-loading bay near the water. On the M/V Sili, everything had to be handed down over the side of the ship and onto the skiff. In addition, there were no Jet Skis to help out this time.

The next day dawned brightly; the rolling sea had settled down. Everyone awakened from the stress of the stormy night and started to move about the deck. It would be several hours before we would see the tops of the palm trees on Swains Island, poking above the perfectly flat horizon. Coffee and food was available in the galley, and everyone seemed content to put the night and the storm behind them. When we saw the first tips of the palm trees, an electric sense of excitement flowed through the entire ship. Jean Michel went forward to the bow to watch the island rise ever so slowly out of the sea. For me, it was an especially great feeling to watch Swains Island come into view like a phoenix, awakening from its slumber and rising anew. A lot had happened in the two years since I had first seen this island rise from the sea, and now here it was, beckoning us again. The Second Expedition was about to land.

On this expedition, however, there was a lot more equipment and, it seemed, people to go ashore. The off-loading process took a few hours, as the M/V *Sili* drifted back and forth offshore of the cut. We had also brought a large, motorized, rubber boat, called a Zodiac, to support dive operations around the island and to assist with unloading. Unfortunately, The Zodiac could not get through the cut in the reef. It was simply too large. This unforeseen problem and the existing wear and tear on the Zodiac, including the condition of its single outboard motor, would make dive operations a little complicated — and interesting. Once all the people and equipment were delivered to the beach, everything had gone quickly, and the M/V *Sili* was released to steam around but stay within radio range.

Setting up Camp. David Jennings, flashing his broad smile, and the advance party were at the water's edge to meet us when we landed (*Picture 10.1*). It was like a great homecoming, there in the far Pacific, on this small and remote island (*Pictures 10.2 and 10.3*). Samoan hospitality and joy infected everyone at the water's edge and set the tone for everything that followed. David and his crew of Samoans had arrived earlier and somehow managed to get a four-wheeler sport vehicle onto the skiff and to the shore. The four-wheeler moved people and equipment to designated areas off the beach and around the Taulaga village area (*Picture 10.4*). David and his crew had been hard at work around the island, tending to a ton of tasks in preparation for the expedition's arrival. An open, covered area for cooking and eating — the galley — had been created under the palm trees just back from the beach (*Picture 10.5*). Large tarps were strung overhead for protection from the sun and frequent rain showers. At the corner of the tarps, 55-gallon blue plastic drums were positioned to collect fresh rainwater. It was a great place to have our meals and to congregate and talk about the events of the day.

Toward the lagoon side of the Taulaga village area, a small building (that I remember more like a screened pavilion) was prepared for the science party and their equipment, including the portable recompression chamber required by NOAA for the diving operation (*Picture 10.6*). (I think it was near this building where, two years before, Captain Wally and I had our early morning chat over coffee that sparked my imagination.) NOAA Dive Guidelines called for a recompression chamber to be within four hours of a dive site. Without the portable chamber, the next closest recompression chamber was in New Zealand, which was many days away (*Picture 10.7*). All diving on this expedition would be by the book. Generators and plentiful fuel were also brought to provide power not only for the recompression chamber but also for computers, monitors, lights, and other sundry needs.

On this trip, Captain Wally stayed aboard the M/V *Sili* as it drifted and steamed around and near the island. As the ship's captain, he was responsible for her. We chatted when I briefly went aboard the M/V *Sili* during dive operations and, of course, on the cruises to and from Swains Island.

Once again, sleeping tents were pitched just back from the beach under the coconut palm trees, and a latrine had been set up nearby. David and his crew also cleared the road/path across the island to the "Old Residence" area, and removed the tangled growth around the cemeteries and the mysterious *tupua*. A number of large trees had come down that blocked the road, but with a little maneuvering, the four-wheeler was able to get around them. The previous expedition was a pretty simple affair in comparison to the range of operations to be undertaken on this expedition. Logistics had to support bigger and more complicated operations on the island, in the lagoon, and offshore. I am sure Alex was proud of how David and his crew prepared the island for this first (and possibly only) comprehensive, science-based, cultural assessment and survey of the island. I know Alex wished he were here with us.

After we finished unloading and ate lunch, David walked with the science party and those new to the island around the village area and along the short trail to the lagoon, explaining the layout of the island and a little of its history. There was still a lot to set up and things to be squared away, but after the warm welcome, everyone seamlessly got to work. Matt and Stephanie found and inspected the dinghy they would use to conduct a side-scan sonar survey of the lagoon (*Picture 10.8*). All of the scuba cylinders brought ashore were placed in an area under the coconut trees, near the beach for easy access, and covered with tarps. Zach Hileman, the NOAA emergency/dive medical technician (*Picture 10.9*), used the four-wheeler to move and set up the large air and oxygen cylinders for the re-compression chamber, and the portable chamber itself, in the pavilion-like building. Zach accompanied the expedition primarily to tend to the recompression chamber and act as a dive master to oversee dive operations. David Jennings, Hans, and Dana were everywhere, ensuring that everyone and everything were settling in. Helping out was Lieutenant Kyle Ryan of the NOAA Corps, whom Hans had recruited primarily to provide logistics support.

Let's Go Diving. I could tell that Jean Michel was anxious to begin diving the coral reef that surrounds Swains Island. The conditions of the corals on this remote atoll, and the species that inhabited them, fascinated him. "Who's in charge now?" he asked me again. "Well, I guess I am in overall charge," I said. "But David's in charge of the island and Hans is in charge of the science party." "Okay, Maestro," Jean Michel declared. "Let's go diving." Given all the things I got him involved in, he began calling me "Maestro" some years before. I was flattered by the impromptu title he bestowed upon me, but I don't think I ever told him so. Such was our relationship.

While Jean Michel, Jim Knowlton, Hans, Dana, Zack, and I discussed the dive operations, members of the science party — Rhonda Suka (geomorphologist), Dave Herdrich (archeologist), and Chris Filimoehala (archeologist) — walked the island to determine where they would begin their investigations. Everything on Swains was a relatively short distance away. You could walk around the lagoon and the entire island, even if you had to bushwhack a little.

All was now set and we were about to get wet for the first time on the expedition. Also for the first time, most of the marine life on the surrounding coral reef would see people blowing strings of bubbles. This was not a reef in the Caribbean!

Chapter 10 Pictures

Picture 10.1. David Jennings on Swains Island.

Picture 10.2. The science party is about to land. Swains Island is in the background.

Picture 10.3. Climbing down into the skiff from the M/V *Sili*.

Picture 10.4. The four-wheeler on Swains Island. Jean Michel, *right*, sits on the vehicle, with the author, *second from right in black shirt*, and David Jennings, *center*.

Picture 10.5. In the galley area, Jean Michel is seated on the box, *left*.

Picture 10.6. The building where the science party worked and slept.

Picture 10.7. Carrying the oxygen cylinders for the recompression chamber across the beach. Hans takes the lead.

Picture 10.8. Preparing the dinghy for the side-scan sonar unit. Matt Lawrence, *right*, and Zachary Hileman.

Picture 10.9. Zachary Hileman, NOAA emergency/dive medical technician.

Chapter 11

Getting Wet

On one occasion, we were pretty far from the M/V Sili and the camp area on the island when we tried to radio the ship to say we were headed back. Either the ship's radio didn't work, no one was listening, or it wasn't turned on. We contacted Zach at the camp, but he couldn't raise the ship, either. We could see the M/V Sili as she came about to get underway and go in the opposite direction, heading out to sea or around the island. By the time we caught up with her, we had plowed around the entire island.

Four of us would do the primary offshore diving on the Second Expedition. Hans and I would dive as buddies, fulfilling NOAA regulations. Jean Michel and Jim Knowlton would dive as the other buddy team. Jim was to video everything of interest while keeping Jean Michel in the frame as much as possible. Although the actual content of the documentary film was yet to be determined, it was clear the film was going to be a Jean Michel Cousteau production. That meant it was critical to capture as much underwater footage of him as possible. The video footage captured on our dives would provide the material for the underwater sequences in the film. Documentaries are just that — they document things — and sometimes there is no real design for the project beforehand. This was one of those cases.

Once we got in the water, I would swim along with Jean Michel and Jim would shadow us with his camera. Hans would then bring up the rear, towing a small surface buoy on a line so that the American Samoan Marine Patrol driver of the Zodiac could follow us from the surface. The four of us had a great deal of diving experience and dived well together as we explored the underwater world off of the island. We were like an experienced team that had been working together for years.

The Dive Plan. In our discussions, Jean Michel hadn't taken long to select the areas where he wanted to dive. From past experience, he had a clear idea of which areas would likely be of most interest. Jean Michel chose three points off the island that represented the highest-energy (or turbulent) edges of the atoll. Swains Island is shaped almost like a distorted rectangle (*Picture 11.1*). The chosen points were like three of the four corners of a bent rectangle. Except for the diving that we had done with the governor two years before, and the diving required from a NOAA ship to place a small monitoring block as part of a Pacific-wide monitoring network, I do not know of any other scuba dives at Swains Island up to that time. If there has been any diving since, I doubt anyone has dived these extremely turbulent areas at the tips of the atoll. Jean Michel's logic was that these areas would likely attract larger predators and be the most interesting, and there would also be plenty of adjacent areas with corals and other species to examine.

Jean Michel's desire to "let's go diving" was easier said than done on this expedition. As it turned out, there were a number of complicated steps to "getting wet." The first step was to communicate by radio with the M/V *Sili* that a diving operation was about to begin. This meant that the ship would have to come close to shore and loiter, waiting for us near the cut through the reef. The crew on the M/V *Sili* would also have to lower the Zodiac with its motor over the starboard side and prepare to receive us on the port side.

Getting Wet Was Not So Simple. Once the ship was in place, we would load the skiff with our gear and tanks and wait, standing on the beach, as Zack would try to communicate with the ship's crew. Each time Zack contacted the ship, he got exasperated and raved on a little about why they couldn't understand him. I remember thinking it was pretty funny. I think the expedition may have been Zack's first time in the faraway South Pacific working with islanders. One day I stood next to Zack and politely reminded him that this was Samoa, not the NOAA Diving Program in Seattle. I said something like, "Don't worry about it, Zack. They probably get it, or we will figure it out when we get aboard." Zack had huffed a little and then just resigned himself to the situation. He knew we could handle it.

The radios themselves were also a problem: from ship to shore, ship to Zodiac, and Zodiac to shore. It was never clear what the problem was at any given time, but it sure hadn't made things any easier. Sometimes it was simply operator error, or dead batteries, or using the wrong radio frequencies. On a couple of occasions, a radio simply had not been turned on. The Samoans didn't worry much about the radios and just did what they thought they should. After all, they could see everything from the deck of the M/V *Sili* and had experience operating without radios. So some mistakes were made, and things always seemed to take a little longer than expected. Jean Michel also got a little frustrated and inquired more than once about who was really in charge. Who was actually in charge depended on the moment. The Samoans just have another way of how they go about their business — and in Samoa, it works.

The next step toward getting wet was to pack ourselves tightly into the skiff with our gear and motor out through the cut in the reef to the M/V *Sili*. Once at the ship, we handed up our tanks and gear, climbed a ladder to the deck, and walked across the ship to other side where the Zodiac was tied (*Picture 11.2*). After a brief chat with Captain Wally and the marine patrol officer who would drive the Zodiac, the final step was to climb down the other side of the ship and load into the Zodiac. It was a heck of a way to go to and from work and a dive site each day. But if it worked for them, it worked for me. I never really figured out why we couldn't just load directly into the Zodiac from the skiff, although the Zodiac was certainly a little more stable alongside of the M/V *Sili*. Thinking about it now, it may simply have been a way to give the M/V *Sili* crew something to do to assist us, and thus be more a part of the expedition.

The Zodiac had seen better days. The burning tropical sun of the South Pacific had seen to that. I remembered that when the Zodiac first arrived in Pago Pago, it had two outboard engines and a sleek-looking center console. On most days, it literally zoomed around Pago Pago harbor as if joy riding. Now the boat had but one cranky outboard left, and its once-black rubber was graying, dried, and brittle. In many places, where stays for lines had been glued to the hull, the stays were gone. The sun had done them in. By the time we would make it back to the M/V *Sili*, one of the inflation cells on the starboard side would deflate and sag. The Zodiac no longer had a Bimini top and was totally exposed to the sun. As Zodiacs go, however, it was roomy and a soft ride, bouncing around offshore of Swains Island. On one occasion, the outboard motor broke down, and we drifted for a few anxious minutes in a very turbulent area near our just-completed dive. Fortunately, the motor kicked over before we had been pushed too close to the rocks.

On another occasion, we were pretty far from the M/V *Sili* and the camp area on the island when we tried to radio the ship to say we were headed back. Either the ship's radio didn't work, no one was listening, or it wasn't turned on. We contacted Zach at the camp, but he couldn't raise the ship, either. We could see the M/V *Sili* as she came about to get underway and go in the opposite direction, heading out to sea or around the island. By the time we caught up with her, we had plowed around the entire island. Apparently, they must have gotten bored on the M/V *Sili*, and rather than standby near the cut, they decided to go for a pleasure cruise around the island. We didn't know where the ship was going, but we had to get aboard to transfer to the skiff. We waved our arms and yelled, although we were too far from the ship to be heard. No one on the M/V *Sili* looked our way or noticed us. By the time we caught up with the ship, pushing the Zodiac for all it was worth, we had completely circled the island and were back offshore of the cut. (It turned out to be the only time I ever went completely around the island offshore.) Maybe we should have just motored to the cut and waited for them to return. I don't think Jean Michel was too happy about chasing the ship that day. After all, who was in charge? All's well that ends well, though, and soon we were back on the island again.

Exploring the Reef. Diving the reef surrounding Swains fully lived up to our expectations (*Picture 11.3*). These dives, combined with the dive two years before, gave me a pretty good appreciation and understanding of this remote coral reef. Everywhere in the clear blue water, the coral reef quickly fell away with no bottom in sight (*Picture 11.4*). Swains Island is on the top of a steep mountain rising up from the seafloor. The reef wraps so tightly around the island that only a few small tidal pools can be seen on the backside of the reef. I don't recall seeing any damaged or diseased corals on my dives at Swains. In places, the large plate corals jutted out and descended like steps into the depths. At one of our dive points, massive waves crashed on the surface as they ran into the island, creating an almost surreal environment. We weren't very deep, maybe only 30 feet or so, and could easily feel the energy and pull of these waves. Jean Michel was right: It was, indeed, a high-energy area. We powered out of their clutches and swam further along the reef face until the pull of the waves dissipated. A few sharks cruised below the turbulent water. It was after this dive that the Zodiac's outboard motor got cranky.

Jean Michel had been looking for a pristine coral reef in the remote South Pacific, and I think he found it at Swains Island (*Picture 11.5*). He also wanted to see and film primary and apex predators. On each of our dives, sharks made an appearance as they hunted along the reef, usually below us. Always inquisitive, Jean Michel frequently paused to peer into nooks and crannies and observe the local inhabitants. Sometimes it looked as if he were carrying on a conversation! Jim faithfully glided along nearby, seeing the reef through the lens of his camera. Only many weeks later did I actually see the footage. Hans had a ringside seat tending his buoy from behind. From his position, he looked like a shepherd or guardian keeping an eye on his flock.

Like other relatively undisturbed coral reefs I have dived in the South Pacific, no single species appeared to be in great abundance on the reef at Swains

Island. Rather, it seemed that many animals shared the reef, with none dominating. The reef at Swains Island may be relatively high in biodiversity. Research by "real" coral reef biologists would be needed to make such a determination with any confidence, though. (I know just enough to be dangerous.)

A Cousteau Film. As I began writing about these dives, I came to realize that I have neither the skill nor words to describe justly the beauty and life we witnessed on the reef at Swains Island. I also didn't possess the scientific expertise to interpret fully what we had seen. Consequently, I decided it would be a disservice to try to do either here. Thankfully, however, the wonderful quality of the reef at Swains Island is shown beautifully in the documentary film made by Jean Michel and Jim Knowlton: *Swains Island — One of the Last Jewels of the Planet.*

For most people, the only way to see a coral reef or begin to understand it is through the camera's eye. So it is for the reef and the life it supports at this tiny place in the remote South Pacific. I recommend this film to anyone interested in life beneath the waves at Swains Island. In addition, Jean Michel's narration in the film provides interesting observations about the corals and animals on the reef. More importantly, though, it reveals how he thinks about life and the planet, which is the reason he made the journey to Swains. I always learn something when I am around him. On many levels, the film does a far better job than I can in conveying the essence and importance of Swains Island and its surrounding reef.

There is one short scene in the film I should clarify, if only just for fun. In the scene, Jean Michel and I swim along the reef, stop, and appear to engage in something profound using sign language and gestures. I am pleased that Jim caught this scene and that it did not wind up on the cutting-room floor. Rather than exchanging something profound about the reef, however, Jean Michel is congratulating me on this day, as it is my 65th birthday. He high fives me and we swim on. He reminded me more than once: What more could I have asked for than to celebrate my birthday under water in the middle of the Pacific Ocean on a pristine coral reef? Thinking back on this now, I recognize how right he was.

After completing two days of wrestling with the Zodiac and diving around the island, we were satisfied with the underwater video Jim had taken. At this point, the expedition's primary diving operations came to an end. After months of preparation, it seemed our dives went by so fast, but we had accomplished what we came to do. Jean Michel and I were soon back aboard the M/V *Sili*, slowly sailing back to Pago Pago to participate in other events. Thankfully, the return cruise was uneventful.

Following our departure, the science party conducted a number of snorkeling and shallow scuba dives from the beach, looking for cultural artifacts, primarily from the later nineteenth and early twentieth centuries. These dives are well documented in the 2013 scientific report, *Unlocking the Secrets of Swains Island: A Maritime Heritage Resources Survey*. The results of these shore dives might be described as slim pickings. Not only are such artifacts hard to find, embedded in coral and debris after so many years, there just wasn't enough time to do a more thorough survey.

Some scuba diving — in addition to the snorkeling Jean Michel, Jim, and I did — was also attempted in the lagoon, as planned. I use the word "attempted" because the conditions in the lagoon, as explained later, were a rude surprise to everyone, including me. The proposition of the lagoon holding unknown artifacts from early Polynesian voyaging days, the primary driver for this expedition, was ultimately thwarted by conditions in the lagoon. These conditions were only hinted at when Jean Michel, Jim, and I snorkeled a small area of the lagoon near the Taulaga village site.

I am certain that islanders must have swum in the lagoon down through the years, and perhaps, even recently. Some probably even surface-dived (holding their breath) and, experiencing something of these conditions, had gone no further. Lastly, I am equally certain that, prior to this expedition, no one had ever attempted to survey the bottom of the lagoon or had tried to scuba dive in it.

Chapter 11 Pictures

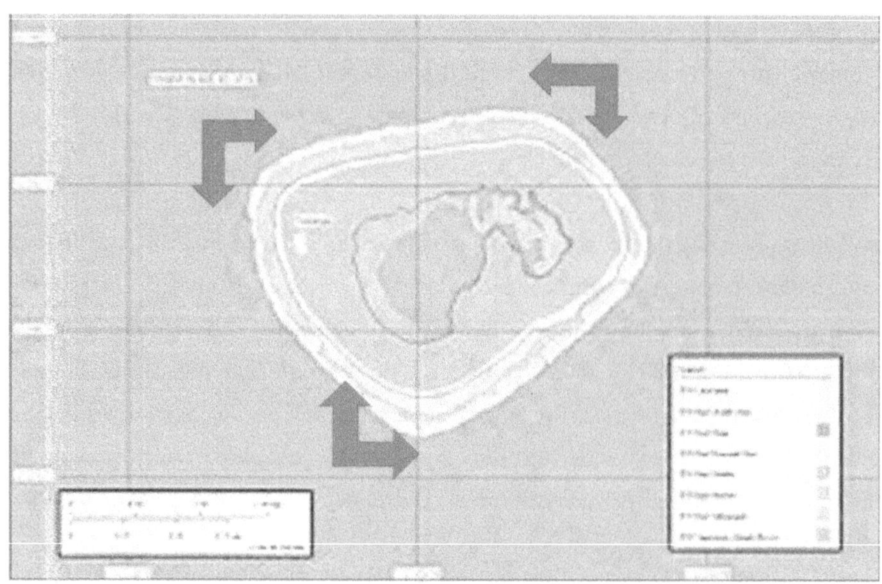

Picture 11.1. Arrows show the approximate locations of the areas we dived.

Picture 11.2. Jean Michel, *left*, next to Jim Knowlton in our Zodiac, alongside the M/V *Sili*.

Picture 11.3. Notice how the reef drops off behind Jean Michel.

Picture 11.4. Swimming along the edge and the rapidly descending slope. Shown here are Jim with his camera, *foreground*, the author, *top left*, and Jean Michel, *top right*.

Picture 11.5. Exploring the reef, Jean Michel encounters a green sea turtle.

Chapter 12

Mysteries of the Tupua
and the Lagoon

The tupua on Swains Island is different from any other known in Polynesia. It may have similarities to tupua elsewhere, but it is different. Research uncovered documented accounts of legends describing Cook Islanders visiting the island and of their "star chart," used to navigate to and from Swains Island. There is now no question that Polynesian voyagers of the past visited Swains Island in their canoes.

The First Expedition hinted that there might be more to do on Swains Island than walk the grounds and ponder sustainable tourism. No one knew why the tupua (a sacred place) was there, in that particular place on the island, or who had constructed it or when, or its purpose. When Captain Wally told me the family story about a former channel into the lagoon and described the "petrifying" properties of the lagoon, my imagination began to connect the dots.

Consequently, we had returned to Swains Island to investigate the mysteries of the tupua and the lagoon, trying to tie it all together. Unraveling the mysteries was the focus of most of the science party's efforts. Their work, in my view, was not that of a typical cultural resource survey. Rather, I like to describe it more along the lines of a forensic investigation, just like a "whodunit" mystery. There were unknowns and questions about what may have taken place and when on Swains Island, in the lagoon, and along the shore surrounding it. The Second Expedition was to uncover any evidence that Swains Island and the lagoon held that could shed light on the $64,000 Question: Does the lagoon hold artifacts of ancient Polynesian voyaging?

Along the way, we also learned much about the island and its cultural landscape, which helped create the background for Jean Michel's documentary film. The film was a priority for me during the expedition. The film's goal was to create interest in Swains Island and American Samoa and to project them onto a larger stage. If artifacts from Polynesian voyaging were indeed found in the lagoon, both Swains Island and American Samoa would be thrust onto the world stage in a big way.

Hans and his colleagues had planned and prepared well for the expedition. The science party consisted of experts in their fields, and most of them had applied their trades in the Pacific. When they left the island, they had exceeded all expectations, especially mine, and shed light on many aspects of the mysteries of the tupua and the lagoon. Their work also resurrected lost knowledge about the role Swains Island had played in the past (*Picture 12.1*).

Had a Channel Existed? Determining if there was a "channel" or a way into the lagoon in the not-too-distant past required an understanding of the geomorphology of the island and the formation of the lagoon itself. Undertaking a geomorphological survey required boring sediment cores, analyzing rock and surface samples, and carefully surveying the lagoon's shoreline to determine how it may have formed and changed over time. Rhonda had to work at full speed to complete the work that was done. (Interestingly, Rhonda was related to the Suka family, which had once lived on Swains Island. I never asked her, but it had to be a great feeling to survey the ground on which generations of family members had once walked.) Findings were uncovered that pointed to how the lagoon was used in the nineteenth and twentieth centuries. Understanding how the lagoon itself functioned, and how the "freshwater lens" beneath the island was created and sustained, was also part of the puzzle to unravel. Water quality samples were taken from the lagoon, along with samples of the biota and algae, and brought back for analysis.

The borings and shovel holes excavated near the tupua and further along the strip of land separating the shallow end of the lagoon from the ocean showed that the terrain consisted primarily of heavy deposits of coral rubble from major deposition events. This end of the lagoon, on the windward side of the island, is exposed to fierce storms and hurricanes. It is entirely different from the leeward side, where we came ashore through the cut. This finding, along with the relic (or fossilized) corals within the lagoon, and together with the family story of a channel, confirmed that the lagoon was once completely open to the sea, and periodically closed and opened again over time. It was perhaps open as recently as 200 years ago.

As in any good whodunit, there were clues to be uncovered everywhere. But the key to solving any mystery is to weave them into the case.

Mystery of the Tupua. Investigations into the tupua turned out to raise more questions than they answered. Archeologists — such as Dave Herdrich, Chris Filimoehala, and others working in the Pacific — still have their hands full trying to uncover what is known about tupua in the Samoan archipelago, nearby islands, and across Polynesia. No written records exist from the Polynesian culture, and all that remains are the physical remnants left behind — platforms, stones, and structures, such as the tupua on Swains Island. Western archeologists have been sketching and describing these cryptic features for more than 150 years, and also trying, whenever possible, to tease out what is known of historic practices, legends, and chants.

Trying to understand the tupua on Swains Island and its purpose started with comparing it to other known tupua generally described in Polynesia. It's like having a large jigsaw puzzle in pieces on the table, and then finding another piece on the floor beneath the table. It may fit here and mean this, or there and mean something entirely different. There is only so much information that can be squeezed from rocks, rubble, and inanimate objects, especially in the relatively short time the science party was on the island.

Upon returning from the expedition, science party members did a scholarly deep dive into the literature that mentions tupua in Polynesia. Their findings go back to the 1840s and are thoroughly reported in *Unlocking the Secrets of Swains Island: A Maritime Heritage Resources Survey*. It is fascinating how they tried to compare the Swains Island tupua to the bits and pieces they found in the literature. If only the Polynesians who built the tupua could speak to us today. Then we might know its true purpose. The conclusion is that more research is needed to answer this question fully. Not an unexpected outcome, but there are some conclusive facts that came from their work.

First is that the tupua on Swains Island is different from any other known throughout Polynesia. It may have similarities to tupua elsewhere, but it is unique *(Picture 12.2)* and so might be its purpose.

Second is that the research uncovered documented accounts of legends describing Cook Islanders visiting Swains Island (known as Olosega) and of their "Star Chart," which they used to navigate to and from the tiny island. There is now no question that Polynesian voyagers in their canoes visited Swains Island in the past.

Third and finally, when taken together, these findings or clues lead the curious detective to conclude it is likely that Polynesian voyagers may have periodically entered the lagoon in the distant past. It follows, then, that the petrified remains of a voyaging canoe, or pieces from one, may well rest at the bottom of the lagoon.

Confounding Mysteries of the Lagoon. Above all other evidence and reasoning that could be uncovered, finding an actual artifact from the Polynesian voyaging era would be the proverbial smoking gun. (Not unlike the detective in an

Agatha Christie novel finally finding the murder weapon.) Only an accurate side-scan sonar survey could reveal if there were any anomalies on the bottom of the lagoon to investigate. Setting up the boat, sonar, and navigation equipment; laying out the survey lines to follow; and testing all systems to do this took some time (*Picture 12.3*). While Matt, Stephanie, and others worked on setting all of this up, Jean Michel and I decided to conduct a snorkel survey of the lagoon offshore of the Taulaga village site to see what we might uncover.

To my knowledge, no one before us had ever snorkeled or surface dived to explore the lagoon. Jim joined us, bringing his video camera to capture what he could. To me it felt like we were the "first" to step into the mysterious lagoon. This certainly could not be true, but my senses were heightened anyway, and I was on full alert to find any evidence that might support any part of the lagoon's story. It turned out to be a very interesting swim.

To get to the lagoon, we hiked the short path from the village, carrying our masks, fins, booties, snorkels, Jim's camera, and a few other things. Walking along the path, if you looked closely at the underbrush, you could see that vegetation had grown over discarded and rusting items from the colonial and copra plantation eras. Later, we learned that the lagoon probably played an important role in transporting coconuts (for making copra) and other things from one side of the island to the other. The copra furnace and storage sheds had been in the Taulaga (*Picture 12.4*) end of the lagoon and close to the cut through the reef. There is no evidence of such facilities having been at the residence end of the lagoon, nor any other way through the fringing reef in that part of the island. In the past, much activity must have taken place on and around the lagoon's shoreline fringes (*Picture 12.5*). Stone piers were located on the other side of the lagoon, and the remains of pilings were found at places along the shore (*Picture 12.6*). Obviously, it was a lot easier to move the heavier items of the day on the lagoon, rather than haul them over the circle road around the lagoon. The lagoon may have been new to us, but many past generations had known it very well. No one during the colonial and plantation eras, however, appeared to have made the possible connection of the lagoon to early Polynesian voyaging; neither had they ever mentioned the real purpose of the tupua.

We had been told that the lagoon's near-shore waters were shallow, but it was a surprise just how shallow it really was when we walked in with our snorkeling gear. As we pushed along toward the small manmade island close to shore, the water was only about two feet deep. There is but one islet in the lagoon; a small tree or two grows on it (*Picture 12.7*). After my first steps, I knew exactly what we were walking on — the remains of a relic coral reef shelf. I had walked on identical surfaces many times. Within a matter of seconds, it was clear to me that the lagoon had been open to the sea in the distant past.

At the tiny islet, we came to the edge of a shelf and a steep face that dropped off much deeper. I surface-dived this face to about 12 feet, where it seemed to flatten out. The bottom beyond was shielded from view by what appeared to be suspended detritus and sedimentary material. It seemed clear that this was the face of a relic reef shoreline (*Picture 12.8*). Swimming along the face, we saw the fossilized remains of corals and large clams, still clinging to the face, and even more scattered at its base where the bottom flattened out. Yes, indeed, the lagoon had definitely been open to the sea, and, ultimately, closed eons ago. I was a little embarrassed about how easy it was to come to this conclusion. Two years ago, it never occurred to me to dip my feet into the lagoon.

The tiny manmade islet at the edge of the coral face, where the deeper water began, probably served some function for the villagers or the copra plantation. The detritus material we encountered was later confirmed to be "green, freshwater algae balls." Later, the science party referred to them as coconut puff balls, because that is exactly what they looked like under water. Jim captured much of what we saw on this snorkel sojourn with his camera (*Picture 12.9*), including the first surprising footage of small freshwater gobies and mollies in this remote lagoon environment otherwise devoid of fish (*Picture 12.10*). There is speculation as to how these fishes might have come to inhabit the lagoon. Scenes of these little fish are included in the documentary film, and Jean Michel's narration explains all that we saw in the lagoon.

Defeated by Coconut Puffs. It was the algae balls — the coconut puffs — and other algae conglomerations that ultimately defeated my simple-minded scheme of easily diving on side-scan sonar targets to identify potential artifacts (*Picture 12.11*). In some places, the algae masses turned out to be almost 16 feet deep, making it impossible to safely penetrate them with our diving equipment. Even so, Matt had given it a go. Having dived previously with Matt in Massachusetts Bay, in dark, cold water under difficult conditions, I knew that if he couldn't get through the algae masses safely then it couldn't be done.

Maybe it was a good thing that I had departed the island with Jean Michel before the side-scan sonar survey was completed and diving in the lagoon was attempted. I have dived in zero visibility many times and would have certainly dived along with Matt and experienced the frustration of attempting to investigate potential targets. Everyone on the expedition was disappointed that the algae masses obfuscated the targets revealed by side-scan sonar. I can think of no way the science party could identify the list of targets on the bottom in those conditions. Artifacts are not just sitting on the bottom saying, "Come find me." I don't think that even dumb luck could have helped. If another expedition is ever to be undertaken, a different approach would have to be devised to penetrate the algae masses of the lagoon. The evidence trail had come to an end.

Nevertheless, we learned a lot about the lagoon — far more than I imagined we would learn when I first looked out onto it two years before. The discovery of symbols of the great age of Polynesian voyaging, however, was not to be. They may still lie hidden under the algal masses in the freshwater lagoon at Swains Island — just waiting to be discovered.

Chapter 12 Pictures

Picture 12.1. The science party on Swains Island. Note the sleeping tents, *upper left,* and eating area under the tarps.

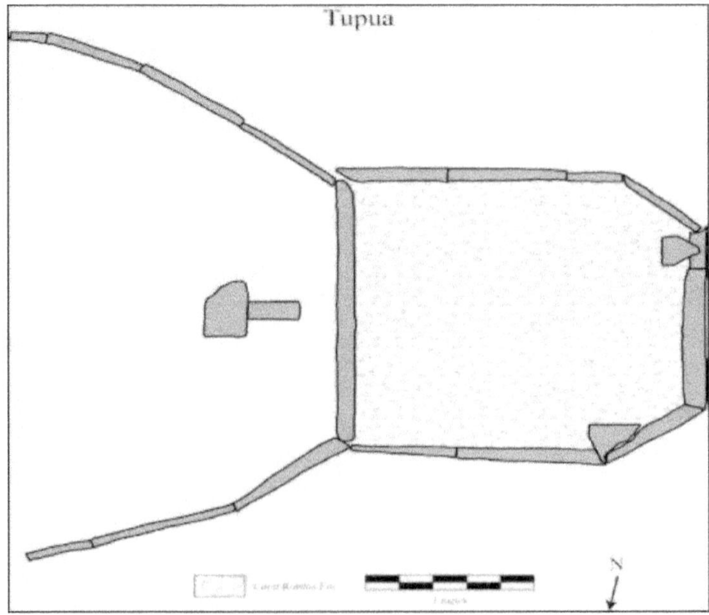

Picture 12.2. Overhead diagram of the tupua.

Picture 12.3. The side-scan sonar survey is underway with Lt. Ryan, *left*, and Matt Lawrence.

Picture 12.4. The copra storage sheds as they appeared during the plantation period.

Picture 12.5. The jungle-fringed edge of the lagoon.

Picture 12.6. A rock pier in the lagoon.

Picture 12.7. The hard work of archeology on Swains Island.

Picture 12.8. The manmade island in the lagoon.

Picture 12.9. Coral rubble in the shallow water of the lagoon near the island.

Picture 12.10. Small mollies swim in the shallow water near the edge of the lagoon.

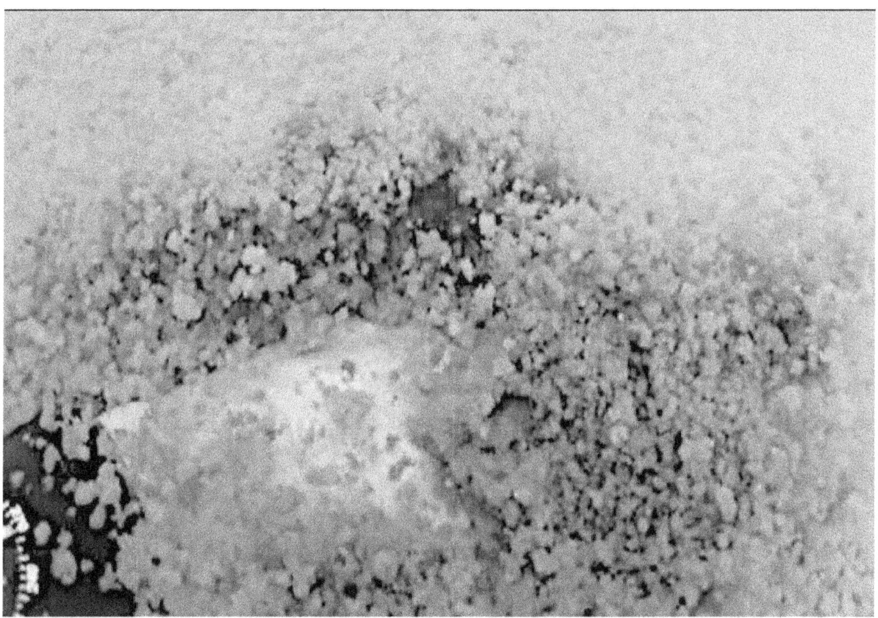

Picture 12.11. The algae "coconut balls" that defeated diving operations in the lagoon. Note the diver's hand.

Chapter 13

Aftermath of the Second Expedition

The story of these expeditions would be unfinished without the telling of what happened after they were completed, especially the Second Expedition. If both expeditions were part of a larger story, then what was that story, and how did the expeditions fit into the effort to ultimately establish a new marine sanctuary in American Samoa? Hopefully, the aftermath of the Second Expedition makes it a little clearer.

The passage back to Pago Pago on the M/V *Sili* was uneventful. I don't think anyone making the early return trip with Jean Michel and me was sorry to get off the ship, nor was the science party and others when they returned five days later. Gene was especially happy to get back and place her feet on solid ground. I don't know if I would have made the journey if I suffered as she does from motion sickness.

The Ocean Center at Pago Pago. The celebration of the one-year anniversary of the Tauese P.F. Sunia Ocean Center in downtown Pago Pago, which Jean

Michel and I had come back to attend, was, in some ways, prophetic. The center represented the efforts of many people. It aimed to bring a new awareness to American Samoans and to people everywhere about how special American Samoa is, and the value of marine protection and education programs to sustain the culture and way of life. The center incorporated many high-tech features to allow the Samoans to look out onto the world, and for the world to look in on the Samoans in return. Jean Michel had been a special guest. To many Samoans, his presence validated the center and so much more. After this first anniversary, the center became central to all visitors coming to American Samoa, as well as a focal point for American Samoans.

In August of 2012, approximately two months before the legislation to create the National Marine Sanctuary of American Samoa was passed that October, the Ocean Center provided the venue for the U.S. Task Force on Coral Reefs' biannual meeting. The center then became the new headquarters of the National Marine Sanctuary of American Samoa. The creation of the epic marine sanctuary was the product of many efforts: people, events including the expeditions, construction of the Ocean Center, the coral reef task force meeting, myriad outreach activities, and the adept political acumen of Governor Tulafono and a few others. The Second Expedition, coming on the heels of the expanded sanctuary and its aftermath, helped to confirm the value and meaning of protecting the far-flung islands of the Pacific.

Before leaving Samoa, Jean Michel and I traveled the seven miles by boat to the small, nearby island of Annu'u, Togiola's birthplace, to visit the island's village and dive its coral reef. The reef and surrounding waters support the primary enterprise on the island — fishing. I remember that it was a drift dive on a bright and windy day, and we both had to swim hard for the boat. Although the reef was in generally good condition, it was not like the pristine one at Swains Island. The reef and waters surrounding Annu'u, like those at Swains Island, were declared Sanctuary Protected Special Areas within the overall new National Marine Sanctuary of American Samoa.

Jim Knowlton Stays Behind. Jean Michel and I departed American Samoa a few days before the return of the science party, Jim Knowlton, and the others. It had not been planned for Jim to stay on the island. I remember that afternoon, standing on the beach with Jim and Jean Michel, before we were to board the skiff to the M/V *Sili* and make the voyage back to Pago Pago. We were discussing the possibility of Jim staying with the science party. It was Jim who had raised the proposition of staying, although it would make his life a little more difficult as well as complicate his connections getting home. To Jim, it was all about the film, and he had been smitten by his experience thus far on the expedition. If Jim stayed, he could video so much more of the science party's efforts, as well as capture more of "the feel" of this remote place. I know it was a no-brainer for me and, I think, for Jean Michel as well. We would find a way to make it work, and so we got into the skiff and departed, leaving Jim on the beach. As I looked back, I saw that Jim was smiling from ear to ear. For a filmmaker like Jim, it was a dream come true — left to wander on a fascinating remote island in the South Pacific, with a mystery to document, and his camera.

Interpreting the Science. As with every scientific expedition, the really hard work begins afterward, and this was certainly the case after the Second Expedition. It would take many months to sift through all the collected data and information. It would require even more time to do additional research to interpret the expedition's findings against the historic background of the islands and the Polynesian culture of the Pacific. Many drafts of the report were passed among the participants and reviewed by others. As lead author, Hans coordinated and knitted together the final report, *Unlocking the Secrets of Swains Island: A Maritime Heritage Resources Survey*, which was published in 2013 as part of the Maritime Heritage Series of NOAA's Office of National Marine Sanctuaries.

I think the report is somewhat of a landmark document on several levels. The report carefully tries to weave together the story of Swains Island and to untangle the mysteries of the tupua and the lagoon. It describes, in a forthright manner, how fragmentary information can be brought together to render a perspective on the historic past, even in such a remote and unknown place as Swains Island. It provides a template or example for others engaging in such work. As a science report, however, it was not widely distributed beyond the immediate communities of interest. It never was intended for the public at large. At the time of this writing, copies may be hard to come by; however, the entire document is available online. The documentary film, though, was intended to reach a wide public audience, perhaps even worldwide.

After returning to Pago Pago, many of us made the long trip home and followed up on the aftermath of our Swains Island experience. When he arrived back in Seattle, Dana Wilkes immediately went to an orthopedist, who quickly informed him that "yes" he had broken his ankle that morning in the dark at the Moana O Sina lodge before boarding the M/V *Sili*. In typical fashion, he mentioned this to me many weeks later. He eventually recovered. Jean Michel and Jim Knowlton immediately began the difficult task of designing the film, editing the footage, and writing the narration. Additional information had to be researched and gathered if a complete story were to be told about Swains Island and its people. I think Jim relied heavily on the Jennings family, especially David Jennings, and on Hans, since he wrote the 2009 background study.

Making the Film. Over the course of developing the film, I discussed its direction and messages with both Jean Michel and Jim. There was always the question if it was going to serve the purposes for which it was being created. By this time, I had worked with Jean Michel on aspects of several other films, and our working relationship seemed effortless. Although Jim and I had worked together only a little in the past, it was the same with him. On a couple of occasions, we met at Jean Michel's Ocean Futures Society headquarters in Santa Barbara, California, to review things. In all honesty, I don't think I brought much to the table. The American Samoa and Swains Island experiences had motivated and focused them both. They are also very good at what they do. I guess what I am most proud of is that I had persuaded them to take on this wing-and-a-prayer assignment, with no pay and little-to-no resources, to make a film of a "maybe story" in such a remote place. They had done the rest.

The premiere, "A Film by Jean Michel Cousteau: *Swains Island — One of the Last Jewels of the Planet*," took place at the Tauese P.F. Sunia Ocean Center in Pago Pago in August 2014, two years after the Second Expedition. Not only did Swains Islanders and Samoans fill the center, but Jean Michel, Jim, and other luminaries — including the world-renowned Sylvia Earle and the Polynesian Navigator himself, Nainoa Thompson — came to Pago Pago to witness the premiere. A fitting complement to the premiere was taking Sylvia, Jean Michel, and Nainoa to dive Fagatele Bay, the tiny American Samoan reef that started it all. The film was shown in many venues and at film festivals around the world, where it won three documentary film awards. I attended a number of these events, sitting on panels with filmmakers and others and examining the making of a documentary that creates a difference (*Picture 13.2*). In the minds of many, Jean Michel and Jim's documentary of the Second Expedition stands out among such films.

One of the film's awards was first prize in the "Cultural Connections: People and the Sea" category at the BLUE Ocean Film Festival in St. Petersburg, Florida, in 2014, following the film's North American premiere (*Picture 13.3*). The Ocean Futures Society owns the film and more information is available on the society's website, as well as on YouTube and the National Marine Sanctuary of American Samoa website.

The Hokule'a Comes to Samoa. In the fall of 2014, Nainoa and the *Hokule'a* sailed to Swains Island (*Picture 13.1*) on her way from Samoa to Tonga during the Worldwide Voyage. This was probably the first time in nearly a thousand years that a Polynesian voyaging canoe landed at Swains Island. The event was broadcast around the world to everyone following the voyage of the *Hokule'a*. Nainoa and his crew went ashore to visit the island and swim in the waters along the reef — part of the new American Samoan National Marine Sanctuary.

I like to think it was my discussions with Nainoa over the years about the Worldwide Voyage and the stories I told him about the Swains Island expeditions that led to the *Hokule'a* visit in 2014. The Worldwide Voyage, primarily sponsored by the Polynesian Voyaging Society, was Nainoa's brainchild. The goal was to circumnavigate the globe with the *Hokule'a* and bring attention to the climatic and ecologic changes affecting the world's oceans — all through the lens of Polynesian values. Such a voyage had never been done. It was an extremely difficult undertaking on every level imaginable, and was years in the making. No Polynesian voyagers are known to have sailed outside of the Pacific Ocean until Nainoa and the *Hokule'a* did so. It was a big thing to visit American Samoa and Swains Island and bring them to the forefront of Polynesia, at least for a time.

The expeditions, along with Jean Michel's documentary and the visit of the *Hokule'a*, all helped to elevate the tiny, remote, uninhabited, and largely unknown Swains Island to the world stage. In today's world of almost unfathomable communication infrastructure, much about the Swains Island story, including some parts of this story, are now available to everyone with a simple internet search. I like to think that much of this material was stimulated by our efforts and the two expeditions.

The run-up to the Swains Island expeditions starting in 2005, and their aftermath were among a series of activities giving momentum and form to the drive to expand the tiny Fagatele Bay National Marine Sanctuary. The role they played, along with other efforts, such as the creation of the Ocean Center, influenced the chain of events that resulted in the National Marine Sanctuary of American Samoa — the end goal of all our efforts. The expeditions gave exposure and recognition not only to Swains Island, but also to American Samoa at a very important time.

***

The expeditions themselves had been serious fun for everyone and served a purpose far greater than Swains Island itself. The path to make them happen had been chaotic at times. In the end, however, Alex finally achieved what he and the entire Jennings family wanted when he first approached me in 2005: to protect their beloved Swains Island. The waters surrounding the island are now protected for all time, and the island and its people's interests are included in a federal program — The National Marine Sanctuary of American Samoa.

The mystery at the bottom of the lagoon remains to be solved, perhaps drawing others to Swains Island. Are artifacts from the great age of Polynesian voyaging hidden in the mists of the Swains Island lagoon? It would be spectacular if they were.

Chapter 13 Pictures

Picture 13.1. The *Hokule'a* at sunset off of Swains Island.

Picture 13.2. *From left*, Jean Michel, the author, and Jim Knowlton at a showing of the film.

Picture 13.3. David Jennings, *left*, and Jim Knowlton accepting the first-place award at the BLUE Ocean Film Festival.

Chapter 14

Epilogue

Who could have predicted that almost 50 years later, the teenage Brooklyn boy who was amazed by the Polynesian "knife and fire" dancers at the 1964 World's Fair, would dance bare-chested with one of those very same young Polynesians in American Samoa in the torch light — thankfully, without the knives and fire. Certainly not me, and not Togiola! I guess such things are what kismet is all about.

At this time, Swains Island remains uninhabited in its remote corner of the South Pacific. It is still visited periodically by the Jennings family and other families who once had made the island their home. A sustainable tourism enterprise still eludes Swains Island and the islanders. The island's waters, however, are now part of the National Marine Sanctuary of American Samoa, and Swains Island is part of the nation's community of marine protected areas. Whether or not this means that more resources will find their way to programming for Swains Island remains to be seen. The size of the tremendously expanded National Marine Sanctuary of American Samoa puts far more demand on the very limited resources formerly allotted to the U.S. territory.

Despite the efforts of many to create the new sanctuary, including the influence of the expeditions to Swains Island, it would not have been possible but for an often-overlooked event in 2009. This pivotal event was President George W. Bush's proclamation that directed:

"The Secretary of Commerce shall initiate the process to add the marine areas of the Rose Island monument to the Fagatele Bay National Marine Sanctuary in accordance with the National Marine Sanctuaries Act (16 U.S.C. 1431 et seq.), including its provision for consultation with an advisory council, to further the protection of the objects identified in this proclamation."

The proclamation directed the government to work to create and expand the tiny Fagatele Bay National Sanctuary, which ultimately resulted in the new National Marine Sanctuary of American Samoa. With a single stroke of the President's pen, many of the formidable barriers to bringing marine protection to American Samoa were removed. The rest is history, including the expeditions to Swains Island themselves. I am sure many don't even remember the proclamation or appreciate the campaign to get the attention of an American president. That is a much bigger story, and for another time.

In American Samoa, the path to the proclamation and, eventually, to the creation of the National Marine Sanctuary of American Samoa took many years, as does the creation of any other marine sanctuary or protected area in the ocean or on land. I think the process actually began when I first stepped foot in the islands, although I was unaware of it at the time. For some unknown reason, and it is almost axiomatic, such community-based processes take on the order of ten years or so to come to fruition. Maybe it's a generational thing or something about the rhythm of building trust. Taken in this light, the Swains Island expeditions were, in fact, important waypoints along the trust-building path.

In my long experience, it takes many people performing many selfless tasks (often seemingly quite small and unrelated) at many levels to build a path, brick by brick, that ultimately leads to the creation of a protected area. As individuals tired along the way, others came forward and carried the bricks; as disillusion set in (and it will), still others stepped forward to reinvigorate the process. The Swains Island expeditions, and many of the other things mentioned in this story, kept the process inspired and moving forward while always demonstrating the trust in, and vision of, the benefits that a National Marine Sanctuary of American Samoa could provide to Samoans and fa'a Samoa. It is not a unique story, but it is a special story to me. In the end, the answers to all things lie within the community itself.

Following is an update on what some of the key individuals mentioned in this book were doing at the time of this writing.

Governor Togiola Tulafono, my dear friend, has since retired, as have I. He still makes his home on his beloved island of Tutuila, not far from Pago Pago, and the village of his youth. Even so, the welfare of greater American Samoa and the Samoan people is always on his mind. He remains a member of the Citizen Advisory Board of the National Marine Sanctuary of American Samoa. At the time of this writing, he is its chairperson. He continues to strive to make the dream come true.

Jean Michel Cousteau still travels the world, making documentary films and preaching about the state of the world's oceans. It is amazing to me how indefatigable he still is, and I wish I had some of his genes. In 2021, his Ocean Futures Society hosted a virtual film festival showing the Swains Island film followed by a live question-and-answer period. Jim Knowlton moderated the Q/A, which featured David Jennings, Hans Von Tilburg, and Jean Michel. Interestingly, when I spoke with Jean Michel before the festival, he was still mulling over how the film could be improved by using a well-known celebrity to introduce the film or perhaps narrate it. I don't think anyone could replace him as the narrator — at least not for my money.

Jim Knowlton continues as an active filmmaker. In 2014, he created Blue Ocean Productions and now works full-time as its owner and director of photography. Of course, he has also branched out into broadcast and media projects. His last big project with Jean Michel was producing and editing *My Father the Captain:*

Jacques-Yves Cousteau, Jean Michel's film about his legendary father.

Alex Jennings continues, with the help of his brother, David, to lead the Swains Island community when he is in Samoa. He remains the representative for Swains Island in the American Samoa *Fono* (governing body) and, as the island's foremost spokesperson, never seems to miss an opportunity to bring attention to it. David continues to make his primary home in Texas, and enjoys his children. His big smile is still unforgettable.

Nainoa Thompson and his team of captains and crew from around the Pacific completed the Worldwide Voyage, called *Malama Honua*, and returned to Hawaii as justly hailed heroes in the Polynesian voyaging tradition. There is no question that it was an epic journey. Nainoa and his family still make their home in the Niu Valley at the Thompson homestead, and Nainoa continues to train new navigators at the Polynesian Voyaging Society's facilities in Honolulu — the home of the *Hokule'a*.

Gene Brighouse stepped down as superintendent of the National Marine Sanctuary of American Samoa, but remains a full-time contractor and works as hard as ever. She splits her time between Western Samoa, American Samoa, and New Zealand.

Lelei Peau left the territorial government and became a federal employee. In 2017, he became the superintendent of the National Marine Sanctuary of American Samoa. Interestingly, Lelei is now known as Atuatasi, his new Samoan high orator's title, as he has been elevated to a higher chief position in the matai hierarchy. I think this makes him the highest-ranking Samoan chief in the U.S. federal service.

Hans Van Tilburg still makes his home on Oahu in Hawaii, although he spent almost a year in the Washington, D.C. area, running the Office of National Marine Sanctuaries Maritime Heritage Program. He is, of course, heavily involved in a number of marine archeology projects across the Pacific. His reputation as one of the region's preeminent marine archeologists continues to grow. He has also become a dedicated cyclist.

Dana Wilkes also remains in the Office of National Marine Sanctuaries and is doing what he does best: overseeing small boats and marine operations. He shows no permanent effects from his once badly broken ankle that he endured during the Second Expedition to Swains Island. He spends a good deal of time working on his cabin in the wilderness near Seattle.

Nancy Daschbach finally retired after a long career in marine conservation and protected area management in the Pacific. She still lives in Hawaii.

Bill Kiene, who early on prompted many of the creative ideas that led to this story, and I continued to work together for many years after he left Samoa, promoting marine protected areas in the Gulf of Mexico and Cuba. He eventually left NOAA and is a sought-after coral-reef scientist currently working where his expertise is needed, including organizations like the United Nations Environment Programme.

Paul Chetirkin, my old campmate and the underwater videographer on the First Expedition, married, started a family, and moved on to a different career.

Stephanie Grandulla returned to the Thunder Bay National Marine Sanctuary in Michigan and continues to work there. She and her husband now own a scuba shop in Alpena, Michigan, where the Thunder Bay sanctuary is head-quartered.

Matt Lawrence and I have dived together on a couple of shipwreck projects along the East Coast since the Second Expedition to Swains Island. Matt gave up the cold waters of New England for the Florida Keys, and now serves as the marine archeologist at the Florida Keys National Marine Sanctuary.

As for the "infamous" M/V *Sili?* Even though time turns slowly in the South

Pacific and American Samoa, it has thankfully and finally caught up with the old ship. Let's just say that she has finally retired and was replaced by a newer, more capable vessel.

***

A Final Note. Lastly, a brief description of my most endearing experience during those years in American Samoa may help sum up why I had to write this story the way I have.

It was sometime in late 2011 or early 2012 at a reception/meeting at the governor's residence in Pago Pago. The residence is a gleaming white, colonial-period wooden building that sits atop a prominent hill overlooking the harbor. Gathered that evening were the governor, his cabinet, and most of the leading American Samoan chiefs on Tutuila. I was there only to attend. At one point, the governor asked the head of the Office of Samoan Affairs to come to the microphone. He was the elderly Paramount Chief Tufele from the Manu'a Islands (or Manu'a tele) almost 70 miles east of Tutuila. As the furthest islands east in the archipelago, these small islands are historically sacred to all Samoans. Tufele's home island was Ofu, one of the three main islands of Manu'a tele.

Tufele was a wise and revered paramount chief. As the head of Samoan affairs, Tufele could be said to be the keeper or guardian of fa'a Samoa in the territorial government. I had first met him more than a decade earlier, and I don't think he was much impressed by me. In his lifetime, and in his post as head of Samoan Affairs, he had dealt with many *Palagis* (non-Samoan white persons) who were ignorant about Samoans but would come to American Samoa to "fix" them, stay a while, and leave only to be replaced by another, and another, and yet another. Early on, I had just been another one of "those people" with whom he had to deal. Something must have changed his views over the years. Perhaps it was just that I didn't seem to go away.

The governor asked me to come to the microphone and stand beside Tufele. Something prepared and expected was about to happen. Tufele began speaking in Samoan and shifted back and forth to English. At first, I didn't know that he was talking about me. Then I looked to the governor, my friend Togiola. His dark eyes glistened and a broad smile almost connected his ears. Tufele, a short man by Samoan standards, made up for his size with a forceful and earnest manner of speaking, as he did that night. He then bestowed upon me the honor of high chief in the Samoan matai from his home village on Ofu and gave me a Samoan name or title: Tui Ali'I. I was dumfounded and, for one of the very few times in my life, at a loss for words. Togiola then returned to the microphone and rescued me from myself. I could utter only a sentence or two of astonished gratitude. Many of the chiefs there came up to me and said: "Now you are one of us." There were smiling faces all around the room. I am sure Togiola was behind it all.

The following summer of 2012, I had a tattoo inscribed around my left arm by a tattoo artist from the Manu'a Islands. It had all been arranged. We sat on mats at the water's edge at Moana O Sina. A cultural representative with a full-body tattoo and several singing women remained there for the three or four hours that the process took. I was told this was the traditional way. The tattoo carries my Samoan name and symbology from the Manu'a tele. In English, Tui Ali'I translates to "supreme high chief."

Things Come Full Circle. Sometimes things do come full circle. That night at Moana O Sina, under the glow of brightly burning torches and many witnesses, Togiola motioned me to a grassy dance area. We would perform a traditional Samoan dance and, as tradition dictated, dance bare-chested in our *lavalavas* (sarongs). Luckily, it was a slow, rhythmic dance, and I easily followed Togiola. Afterward, people asked me how I knew Samoan dances — I didn't. The flickering light of the torches had masked a lot. Later that night, a few Samoan men came to me with leaves they had gathered to wrap around my tattooed arm to ease the pain and heal it while I slept. I did as they said, and the next morning, all was well (*Picture 14.1*).

Who could have predicted that almost 50 years later, the teenage Brooklyn boy who was amazed by the Polynesian "knife and fire" dancers at the 1964 World's

Fair, would dance bare-chested with one of those very same young Polynesians in American Samoa in the torch light — thankfully, without the knives and fire. Certainly not me, and not Togiola! I guess such things are what kismet is all about.

Picture 14.1. The author, *right*, wearing his "chief's garland" and showing the Tui Ali'I tattoo, with the tattoo artist from Manu'a tele.

The End — *Fa'a Samoa*

About the Author

Daniel J. Basta was the Director of the Office of National Marine Sanctuaries in the National Oceanic and Atmospheric Administration (NOAA), within the U.S. Department of Commerce, from 1999 to 2016, when he retired after 37 years of government service. He was a member of the Senior Executive Service (SES) for more than two decades and had an extraordinary career both inside and outside of government. He is well known as an explorer, adventurer, and master diver who has traveled the world. Since retirement, he has become an author, spinning the tales of his adventures, sometimes to the wildest places, and the colorful people he meets. This is his third book.